A GUIDE
TO PHYSICS
PROBLEMS
part 1
Mechanics, Relativity, and Electrodynamics

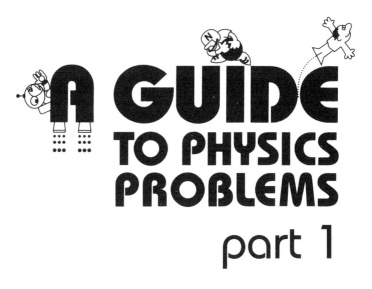

A GUIDE TO PHYSICS PROBLEMS

part 1

Mechanics, Relativity, and Electrodynamics

Sidney B. Cahn
Boris E. Nadgorny

State University of New York at Stony Brook
Stony Brook, New York

PLENUM PRESS · NEW YORK AND LONDON

Library of Congress Cataloging in Publication Data

Cahn, Sidney B.
   A guide to physics problems / Sidney B. Cahn and Boris E. Nadgorny.
      p.   cm.
   Includes bibliographical references and index.
   Contents: pt. 1. Mechanics, relativity, and electrodynamics.
   ISBN 0-306-44679-0
   1. Physics—Problems, exercises, etc. I. Nadgorny, Boris E. II. Title.
QC32.C25 1994   94-5210
530′.076—dc20   CIP

Front cover design by Vladimir Gitt.
Back cover illustration by Yair Minsky.

ISBN 0-306-44679-0

©1994 Plenum Press, New York
A Division of Plenum Publishing Corporation
233 Spring Street, New York, N.Y. 10013

Printed in the United States of America

Foreword

For many graduate students of physics the written qualifying examination is the last and one of the most important of the hundreds of grueling examinations that they have had to take in their career. I remember vividly my own experience in 1947 at the University of Chicago. After the qualifying examination, I knew I was finally free from all future examinations, and that generated a wonderful feeling of liberation and relief.

Be that as it may, the written qualifying examination does serve a useful purpose, both for the faculty and for the students themselves. That is why so many universities give these exams year after year in all parts of the world.

Sidney Cahn and Boris Nadgorny have energetically collected and presented solutions to about 140 problems from the exams at many universities in the United States and one university in Russia, the Moscow Institute of Physics and Technology. Some of the problems are quite easy, others quite tough; some are routine, others ingenious. Sampling them I am reminded of the *tripos* questions of Cambridge University that I had spent so many hours on when I was an undergraduate student in China during the years 1938–1942, studying such books as Whittaker's *Analytical Dynamics*, Whittaker and Watson's *Modern Analysis*, Hardy's *Pure Mathematics*, and Jeans' *Electricity and Magnetism*.

It is perhaps interesting to the readers of this volume to note that the famous Stokes' theorem, so important to modern differential geometry and to physics, first appeared in public as problem No. 8 of the Smith Prize Examination of 1854. Stokes was the examiner and Maxwell was one of the takers of the examination. That Maxwell was impressed with this theorem, and made extensive use of it in 1856 in the first of his epoch-making series of papers that led to Maxwell's equations, is obvious from his papers and from his *A Treatise on Electricity and Magnetism* (1873). Maybe a hundred years from now somebody will remember one of the problems of the present collection?

C.N. Yang

Stony Brook

Preface

The written qualifying examination, a little publicized requirement of graduate physics programs in most universities, brings some excitement to the generally dull life of the graduate student. While undergoing this ordeal ourselves, we were reminded of the initiation ceremonies into certain strict monastic orders, designed to cause the novices enough pain to make them consider their vocation seriously. However, as the memory of the ghastly experience grows dim, our attitudes are gradually changing, and we now may agree that these exams help assure a minimal level of general physics knowledge necessary for performing successful research. Still, the affair is rather stressful, sometimes more a test of character than of knowledge (see Figure P.1). Perhaps it is the veteran's memory of this searing, yet formative experience that preserves the Institution of the Qualifying Exam.

Some schools do not have written exams, for instance: *Brown, Cal-Tech, Cornell, Harvard, UT Austin, Univ. of Toronto, Yale*. However, the majority do administer them and do so in a more or less standard form, though, the level of difficulty of the problems, their style, etc., may differ substantially from school to school. Our main purpose in publishing this book — apart from the obvious one to become rich and famous — is to assemble, as far as possible, a universal set of problems that the graduate student should be able to solve in order to feel comfortable and confident at the exam. Some books containing exam problems from particular universities (*Chicago, Berkeley, Princeton*) have been published; however, this is the first book to contain problems from different American schools, and for comparison, problems from *Moscow Phys-Tech*, one of the leading Russian universities.

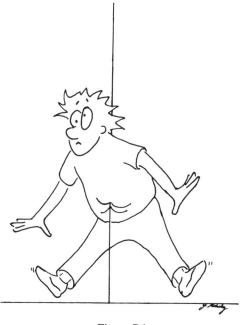

Figure P.1

Hapless Physicist Impaled on his own Delta Function
(Demonstrating the Perils of Insufficient Theoretical Rigor)

The other goal of the book is much more complicated and only partly realized: to allow comparison of problems from different schools in terms of breadth of material, style, difficulty, etc. This would have required analysis of a greater number of problems than we were able to include, and the use of approximately the same number of problems from each department (we had only a few problems from some universities and hundreds from others). We were much more concerned to present problems that would cover as much material as possible. We should note in this regard that the exams with the most difficult problems to solve are not necessarily the most difficult to pass — that depends on the number of problems that have to be solved, the amount of time given for each problem, and the way in which the problems are graded. We have not attempted to present such information, but we wish to point out that it is an important consideration in the selection of a graduate school and well worth investigating.

Quite often the written exam consists of two parts: the first part, covering "fundamental" physics, usually includes classical mechanics, electrodynamics, quantum mechanics, statistical physics and thermodynamics, and sometimes special relativity and optics; the second part, containing "modern" physics, includes nuclear, atomic, elementary particle, and solid state physics, and sometimes general relativity and astrophysics. The scope and difficulty of the second part vary too much from school to school to allow generalization, and we will only deal with the first part. The problems will appear in two volumes: Part 1 — Mechanics, Relativity, and Electrodynamics, and Part 2 — Quantum Mechanics and Statistical Physics.

While reviewing the material submitted to us, we were not surprised to find that often the same problems, maybe in slightly different formulations, were part of the exams at several schools. For these problems, we have noted the name of the school whose particular version we solved next to the name we assigned to the problem, followed by the name or names of schools whose exams contained variants of the problem. If only part of the problem was used at a different school, we have indicated which one. We have also tried to establish a balance between standard problems that are popular with many physics departments and more original problems, some of which we believe have never been published. Many of the standard problems used in the exams have been published previously. In most cases, though, it is difficult to determine when the problem was first presented; almost as difficult as it is to track down the origin of a fairy tale. However, when we could refer to a standard textbook where the problem may be found, we have done so. Although it may be boring to solve a lot of the standard problems, it is worthwhile – usually they comprise more than half of all the problems given in the exams. We have to acknowledge grudgingly that all errors in the formulation of the problems and solutions are the sole responsibility of the authors. We have tried to provide solutions that are as detailed as possible and not skip calculations even if they are not difficult. We cannot claim that we have the best possible solutions and inevitably there must be some errors, so we would welcome any comments or alternative solutions from the reader.

We were encouraged by the response from most of the schools that we approached, which furnished us with problems for inclusion in this book. We would like to take this opportunity to thank the Physics Departments at Boston University (*Boston*), University of Colorado at Boulder (*Colorado*), Columbia University [Applied Physics] (*Columbia*), University of Maryland (*Maryland*), Massachusetts Institute of Technology (*MIT*), University of Michigan (*Michigan*), Michigan State University (*Michigan State*), Michigan Technological University (*Michigan Tech*), Princeton University (*Princeton*), Rutgers University (*Rutgers*), Stanford University (*Stanford*), State University of New York at Stony Brook (*Stony Brook*), University of Wisconsin (*Wisconsin-Madison*). The problems from Moscow Institute of Physics and Technology (*Moscow Phys-Tech*) came from different sources

— none from graduate qualifying exams, rather from undergraduate exams, oral exams, and magazines (Kvant). A few were published before, in a book containing a lot of interesting problems from Moscow Phys-Tech, but most were compiled by the authors. We wish to thank Emmanuel I. Rashba, one of the authors of that book, for his advice. We realize that there are many schools which are not represented here, and we welcome any submissions for Part 2 of this project.

It is our pleasure to thank many members of the Department of Physics at Stony Brook for their encouragement during the writing of this book, especially Andrew Jackson, Peter Kahn and Gene Sprouse, as well as Kirk McDonald of Princeton. We are indebted to Chen Ning Yang, who agreed to write the foreword for this book. We are grateful to: Dmitrii Averin, Fabian Essler, Gerald Gwinner, Sergey Panitkin, Babak Razzaghe-Ashrafi, Sergey Shokhor and Henry Silsbee for numerous discussions of problems and many useful suggestions, and especially to Bas Peeters, who read most of the manuscript; and to Michael Bershadsky, Claudio Corianò, and Sergey Tolpygo for contributing some of the problems. One of the authors (B.N.) wishes to thank the students at Oxford University and Oxford's Student Union for their invaluable help without which this book might not have been written. Finally, we would like to thank Vladimir Gitt and Yair Minsky for drawing the humorous pictures, and Susan Knapp for typing part of the manuscript.

Sidney B. Cahn

Stony Brook   Boris E. Nadgorny

Textbooks Used in the Preparation of This Volume

Chapter 1

An exhaustive bibliography may be found in Goldstein.

1) Landau, L.D., and Lifshitz, E.M., *Mechanics,* Volume 1 of *Course of Theoretical Physics,* 3rd ed., Elmsford, New York: Pergamon Press, 1976

2) Goldstein, H., *Classical Mechanics,* 2nd ed., Reading, MA: Addison-Wesley, 1981

3) Barger, V.D., and Olsson, M.G., *Classical Mechanics, A Modern Perspective,* New York: McGraw-Hill, 1973

4) Routh, E., *Dynamics of a System of Rigid Bodies,* New York: Dover, 1960

5) Arnold, V. I., *Mathematical Methods of Classical Mechanics,* 2nd ed., New York: Springer-Verlag, 1978

6) Landau, L.D., and Lifshitz, E.M., *Fluid Mechanics,* Volume 6 of *Course of Theoretical Physics,* 2nd ed., Elmsford, New York: Pergamon Press, 1987

Chapter 2

1) Taylor, E.F., and Wheeler, J.A., *Spacetime Physics,* San Francisco, California: W.H. Freeman and Company, 1966

2) Landau, L.D., and Lifshitz, E.M., *Classical Theory of Fields,* Volume 2 of *Course of Theoretical Physics,* 4th ed., Elmsford, New York: Pergamon Press, 1975

3) Halzen, F., and Martin, A., *Quarks and Leptons,* New York: John Wiley & Sons, Inc., 1984

4) Jackson, J.D., *Classical Electrodynamics,* New York: John Wiley & Sons, Inc., 1975

Chapter 3

An exhaustive bibliography may be found in Jackson.

1) Jackson, J.D., *Classical Electrodynamics,* New York: John Wiley & Sons, Inc., 1975

2) Landau, L.D., and Lifshitz, E.M., *Classical Theory of Fields,* Volume 2 of *Course of Theoretical Physics,* 4th ed., Elmsford, New York: Pergamon Press, 1975

3) Landau, L.D., Lifshitz, E.M., and Pitaevskiĭ, L.P., *Electrodynamics of Continuous Media,* Volume 8 of *Course of Theoretical Physics,* 2nd ed., Elmsford, New York: Pergamon Press, 1984

4) Panofsky, W., and Philips, M., *Classical Electricity and Magnetism,* 2nd ed., Reading, MA: Addison-Wesley, 1962

5) Marion, J.B., and Heald, M.A., *Classical Electromagnetic Radiation,* 2nd ed., New York: Academic Press, 1980

6) Smythe, W.R., *Static and Dynamic Electricity,* 3rd ed., New York: Hemisphere Publishing Corp., 1989

Note: CGS units are uniformly used in Chapter 3 for the purpose of consistency, even if the original problem was given in other units.

PART I: PROBLEMS

PART II: SOLUTIONS

PART III: APPENDIXES

A GUIDE
TO PHYSICS
PROBLEMS
part 1
Mechanics, Relativity,
and Electrodynamics

PROBLEMS

Mechanics

1.1 Falling Chain (MIT, Stanford)

A chain of mass M and length L is suspended vertically with its lower end touching a scale. The chain is released and falls onto the scale. What is the reading of the scale when a length x of the chain has fallen? Neglect the size of the individual links.

1.2 Cat and Mouse Tug of War (Moscow Phys-Tech, MIT)

A rope is wrapped around a fixed cylinder as shown in Figure P.1.2. There is friction between the rope and the cylinder, with a coefficient of friction

Figure **P.1.2**

μ; the angle $\theta_0 = \pi/3$ defines the arc of the cylinder covered by the rope. The rope is much thinner than the cylinder. A cat is pulling on one end of the rope with a force F while 10 mice can just barely prevent it from sliding by applying a total force $f = F/10$.

a) Does the minimum force necessary to prevent the rope from sliding depend on the diameter of the cylinder?

b) Through what minimum angle θ_1 about the cylinder should one mouse wrap the rope in order to prevent the cat from winning the game of tug of war?

1.3 Cube Bouncing off Wall (Moscow Phys-Tech)

An elastic cube sliding without friction along a horizontal floor hits a vertical wall with one of its faces parallel to the wall. The coefficient of friction between the wall and the cube is μ. The angle between the direction of the velocity $\mathbf{v}$ of the cube and the wall is α. What will this angle be after the collision (see Figure P.1.3 for a bird's-eye view of the collision)?

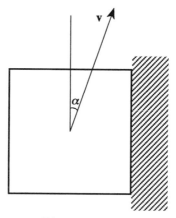

Figure   **P.1.3**

1.4 Cue-Struck Billiard Ball (Rutgers, Moscow Phys-Tech, Wisconsin-Madison (a))

Consider a homogeneous billiard ball of mass m and radius R that moves on a horizontal table. Gravity acts downward. The coefficient of kinetic friction between the ball and the table is μ, and you are to assume that there is no work done by friction for pure rolling motion. At time $t = 0$, the ball is struck with a cue, which delivers a force pulse of short duration.

Its impulse is

$$\mathbf{P} = \int\limits_{-\epsilon}^{+\epsilon} \mathbf{F}(t)\, dt$$

a) The point of contact between the cue and the ball is at the "equator" and the direction of the force is toward the center of the ball. Calculate the time at which pure rolling motion begins. What is the final speed of the center of mass of the ball?

b) At what height h above the center must the cue strike the ball so that rolling motion starts immediately (see Figure P.1.4)?

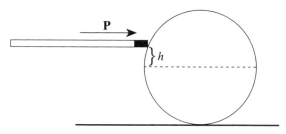

Figure   P.1.4

1.5   Stability on Rotating Rollers (Princeton)

A uniform thin rigid rod of mass M is supported by two rotating rollers whose axes are separated by a fixed distance a. The rod is initially placed at rest asymmetrically, as shown in Figure P.1.5a.

a) Assume that the rollers rotate in opposite directions. The coefficient of kinetic friction between the bar and the rollers is μ. Write the equation of motion of the bar and solve for the displacement $x(t)$ of the center C of the bar from roller 1, assuming $x(0) = x_0$ and $\dot{x}(0) = 0$.

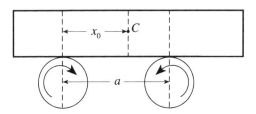

Figure   P.1.5a

b) Now consider the case in which the directions of rotation of the rollers are reversed, as shown in Figure P.1.5b. Calculate the displacement $x(t)$ again, assuming $x(0) = x_0$ and $\dot{x}(0) = 0$.

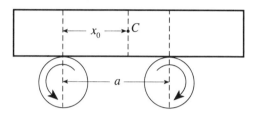

Figure P.1.5b

1.6   Swan and Crawfish (Moscow Phys-Tech)

Two movers, Swan and Crawfish, from Swan, Crawfish, and Pike, Inc., must move a long, low, and narrow dresser along a rough surface with a coefficient of friction $\mu = 0.5$ (see Figure P.1.6). The mass M of the dresser is 150 kg. Swan can apply a maximum force of 700 N, and Crawfish 350 N.

Figure   P.1.6

Obviously, together they can move the dresser; however, each of them insists on his own way of moving the darn thing, and they cannot agree. Show that by using his own method, each of them can move the dresser alone. What are these methods?

Hint: The names in the problem are not quite coincidental, and the two methods are natural for Swan and Crawfish.

1.7   Mud from Tire (Stony Brook)

A car is stuck in the mud. In his efforts to move the car, the driver splashes mud from the rim of a tire of radius R spinning at a speed v, where $v^2 > gR$. Neglecting the resistance of the air, show that no mud can rise higher than a height $R + v^2/2g + gR^2/2v^2$ above the ground.

1.8   Car down Ramp up Loop (Stony Brook)

A car slides without friction down a ramp described by a height function $h(x)$, which is smooth and monotonically decreasing as x increases from 0 to L. The ramp is followed by a loop of radius R. Gravitational acceleration is a constant g in the negative h direction (see Figure P.1.8).

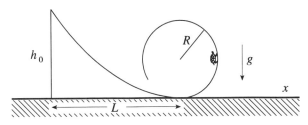

Figure   **P.1.8**

a) If the velocity is zero when $x = 0$, what is the minimum height $h_0 = h(0)$ such that the car goes around the loop, never leaving the track?

b) Consider the motion in the interval $0 < x < L$, before the loop. Assuming that the car always stays on the track, show that the velocity in the x direction is related to the height as

$$\dot{x} = \sqrt{\frac{2g\,[h_0 - h(x)]}{1 + (dh/dx)^2}}$$

c) In the particular case that $h(x) = h_0\,[1 - \sin(\pi x/2L)]$ show that the time elapsed in going down the ramp from $(0, h_0)$ to $(L, 0)$ can be expressed as $T = (L/\sqrt{gh_0})f(a)$, where $a \equiv \pi h_0/2L$, and write $f(a)$ as a definite integral. Evaluate the integral in the limiting case $h_0 \gg L$, and discuss the meaning of your answer.

1.9   Pulling Strings (MIT)

A mass m is attached to the end of a string. The mass moves on a frictionless table, and the string passes through a hole in the table (see Figure P.1.9), under which someone is pulling on the string to make it taut at all times. Initially, the mass moves in a circle, with kinetic energy E_0. The string is then slowly pulled, until the radius of the circle is halved. How much work was done?

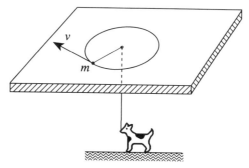

Figure   **P.1.9**

1.10   Thru-Earth Train (Stony Brook, Boston (a), Wisconsin-Madison (a))

A straight tunnel is dug from New York to San Francisco, a distance of 5000 kilometers measured along the surface. A car rolling on steel rails is released from rest at New York, and rolls through the tunnel to San Francisco (see Figure P.1.10).

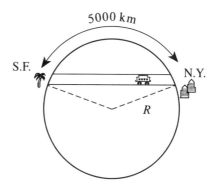

Figure   **P.1.10**

a) Neglecting friction and also the rotation of the Earth, how long does it take to get there? Take the gravitational acceleration $g = 980$ cm/s^2 and the radius of the Earth $R = 6400$ km.

b) Suppose there is now friction proportional to the square of the velocity (but still ignoring the rotation of the Earth). What is the equation for the phase space trajectory? Introduce suitable symbols for the constant of proportionality and for the mass of the car, and also draw a sketch.

c) We now consider the effects of rotation. Estimate the magnitude of the centrifugal and Coriolis forces relative to the gravitational force (ignore friction). Take New York and San Francisco to be of equal latitude (approximately 40° North).

1.11 String Oscillations (Moscow Phys-Tech)

The frequency of oscillation of a string depends on its length L, the force applied to its ends T, and the linear mass density ρ. Using dimensional analysis, find this dependence.

1.12 Hovering Helicopter (Moscow Phys-Tech)

A helicopter needs a minimum of a 100 hp engine to hover (1 hp = 746 W). Estimate the minimum power necessary to hover for the motor of a 10 times reduced model of this helicopter (assuming that it is made of the same materials).

1.13 Astronaut Tether (Moscow Phys-Tech, Michigan)

An astronaut of total mass 110 kg was doing an EVA (spacewalk, see Figure P.1.13) when his jetpack failed. He realized that his only connection to

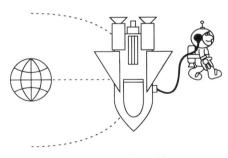

Figure **P.1.13**

the spaceship was by the communication wire of length $L = 100$ m. It can support a tension of only 5 N before parting. Estimate if that is enough to keep him from drifting away from the spaceship. Assume that the height of the orbit is negligible compared to the Earth's radius (R = 6400 km). Assume also that the astronaut and the spaceship remain on a ray projecting from the Earth's center with the astronaut further away from the Earth.

1.14   Spiral Orbit (MIT)

A particle moves in two dimensions under the influence of a central force determined by the potential $V(r) = \alpha r^p + \beta r^q$. Find the powers p and q which make it possible to achieve a spiral orbit of the form $r = c\theta^2$, with c a constant.

1.15   Central Force with Origin on Circle (MIT, Michigan State)

A particle of mass m moves in a circular orbit of radius R under the influence of a central force $F(r)$. The center of force C lies at a point on the circle (see Figure P.1.15). What is the force law?

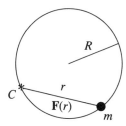

Figure **P.1.15**

1.16   Central Force Orbit (Princeton)

a) Find the central force which results in the following orbit for a particle:

$$r = a(1 + \cos\theta)$$

b) A particle of mass m is acted on by an attractive force whose potential is given by $U \propto r^{-4}$. Find the total cross section for capture of the particle coming from infinity with an initial velocity v_∞.

1.17 Dumbbell Satellite (Maryland, MIT, Michigan State)

Automatic stabilization of the orientation of orbiting satellites utilizes the torque from the Earth's gravitational pull on a non-spherical satellite in a circular orbit of radius R. Consider a dumbbell-shaped satellite consisting of two point masses of mass m connected by a massless rod of length $2l$, much less than R where the rod lies in the plane of the orbit (see Figure P.1.17). The orientation of the satellite relative to the direction toward the Earth is measured by angle θ.

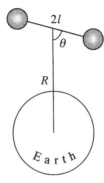

Figure **P.1.17**

a) Determine the value of θ for the stable orientation of the satellite.
b) Show that the angular frequency of small-angle oscillations of the satellite about its stable orientation is $\sqrt{3}$ times the orbital angular velocity of the satellite.

1.18 Yukawa Force Orbit (Stony Brook)

A particle of mass m moves in a circle of radius R under the influence of a central attractive force

$$F = -\frac{K}{r^2}e^{-r/a}$$

a) Determine the conditions on the constant a such that the circular motion will be stable.
b) Compute the frequency of small radial oscillations about this circular motion.

1.19   Particle Colliding with Reflecting Walls (Stanford)

Consider a particle of mass m moving in two dimensions between two perfectly reflecting walls which intersect at an angle χ at the origin (see Figure P.1.19). Assume that when the particle is reflected, its speed is unchanged and its angle of incidence equals its angle of reflection. The particle is attracted to the origin by a potential $U(r) = -c/r^3$, where c is some constant.

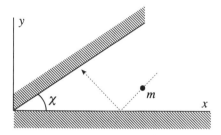

Figure **P.1.19**

Now start the particle at a distance R from the origin on the x-axis with a velocity vector $\mathbf{V} = (V_x, V_y)$. Assume $V_y \neq 0$, $V_x < 0$.

a) Determine the equation for distance of closest approach to the origin.
b) Under what conditions will the particle reach the origin?
c) Under what circumstance will it escape to infinity?

1.20   Earth–Comet Encounter (Princeton)

Find the maximum time a comet (C) of mass m following a parabolic trajectory around the Sun (S) can spend within the orbit of the Earth (E). Assume that the Earth's orbit is circular and in the same plane as that of the comet (see Figure P.1.20).

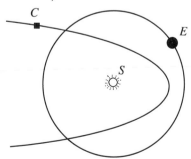

Figure **P.1.20**

1.21   Neutron Scattering (Moscow Phys-Tech)

Neutrons can easily penetrate thick lead partitions but are absorbed much more efficiently in water or in other materials with high hydrogen content. Employing only classical mechanical arguments, give an explanation of this effect (see Figure P.1.21).

Figure **P.1.21**

1.22   Collision of Mass–Spring System (MIT)

A mass m_1, with initial velocity V_0, strikes a mass–spring system m_2, initially at rest but able to recoil. The spring is massless with spring constant k (see Figure P.1.22). There is no friction.

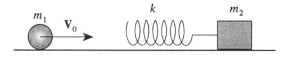

Figure **P.1.22**

a) What is the maximum compression of the spring?
b) If, long after the collision, both objects travel in the same direction, what are the final velocities V_1 and V_2 of m_1 and m_2, respectively?

1.23   Double Collision of Mass–Spring System (Moscow Phys-Tech)

A ball of mass M moving with velocity V_0 on a frictionless plane strikes the first of two identical balls, each of mass $m = 2$ kg, connected by a massless spring with spring constant $k = 1$ kg/s^2 (see Figure P.1.23). Consider the collision to be central and elastic and essentially instantaneous.

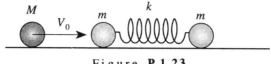

Figure **P.1.23**

a) Find the minimum value of the mass M for the incident ball to strike the system of two balls again.
b) How much time will elapse between the two collisions?

1.24   Small Particle in Bowl (Stony Brook)

A small particle of mass m slides without friction on the inside of a hemispherical bowl, of radius R, that has its axis parallel to the gravitational field g. Use the polar angle θ (see Figure P.1.24) and the azimuthal angle φ to describe the location of the particle (which is to be treated as a point particle).

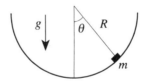

Figure **P.1.24**

a) Write the Lagrangian for the motion.
b) Determine formulas for the generalized momenta p_θ and p_φ.
c) Write the Hamiltonian for the motion.
d) Develop Hamilton's equations for the motion.
e) Combine the equations so as to produce one second order differential equation for θ as a function of time.
f) If $\theta = \theta_0$ and $\dot\theta = 0$, independent of time, calculate the velocity (magnitude and direction).
g) If at $t = 0$, $\theta = \theta_0$, $\dot\theta = 0$, and $\dot\varphi = 0$, calculate the maximum speed at later times.

1.25   Fast Particle in Bowl (Boston)

A particle constrained to move on a smooth spherical surface of radius R is projected horizontally from a point at the level of the center so that its an-

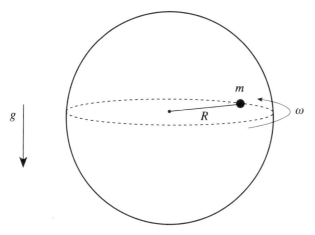

Figure **P.1.25**

gular velocity relative to the axis is ω (see Figure P.1.25). If $\omega^2 R \gg g$, show that its maximum depth z below the level of the center is approximately

$$z \cong \frac{2g}{\omega^2} \sin^2 \left(\frac{\omega t}{2} \right)$$

1.26 Mass Orbiting on Table (Stony Brook, Princeton, Maryland, Michigan)

A particle of mass M is constrained to move on a horizontal plane. A second particle, of mass m, is constrained to a vertical line. The two particles are connected by a massless string which passes through a hole in the plane (see Figure P.1.26). The motion is frictionless.

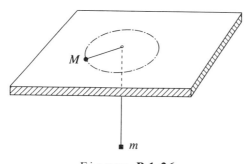

Figure **P.1.26**

a) Find the Lagrangian of the system and derive the equations of motion.
b) Show that the orbit is stable with respect to small changes in the radius, and find the frequency of small oscillations.

1.27 Falling Chimney (Boston, Chicago)

A tall, slender, cylindrical brick chimney of height L is slightly perturbed from its vertical equilibrium position so that it topples over, rotating rigidly around its base B until it breaks at a point P. Show that the most likely value for the distance l of P from B is $L/3$. Assume that the chimney breaks because the torque is too great and the chimney bends and snaps (see Figure P.1.27).

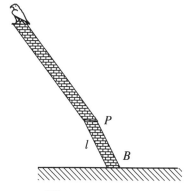

Figure **P.1.27**

1.28 Sliding Ladder (Princeton, Rutgers, Boston)

A ladder of mass m and length $2l$ stands against a frictionless wall with its feet on a frictionless floor. If it is let go with initial angle α_0, what will be the angle when the ladder loses contact with the wall (see Figure P.1.28)?

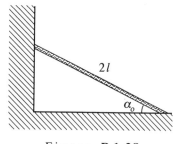

Figure **P.1.28**

1.29   Unwinding String (MIT, Maryland (a,b), Chicago (a,b))

A point mass m is attached to a long, massless thin cord whose other end is attached to a fixed cylinder of radius R. Initially, the cord is snugly and completely wound up around a circular cross section of the cylinder, so that the mass touches the cylinder. No external forces are acting, except for an impulse at $t = 0$ directed radially outward to give the mass m an initial velocity of magnitude v_0. This starts the mass unwinding (see Figure P.1.29a). The point P is the initial position of the mass, and Q denotes the instantaneous contact point between the cord and the cylinder.

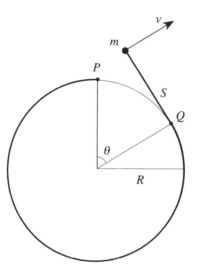

Figure **P.1.29a**

a) Find the Lagrangian and equation of motion in terms of the generalized coordinate θ as a function of time, satisfying the initial conditions.

b) Using the above solution, find the angular momentum of the mass about the center of the cylinder. Is angular momentum conserved? Why? Is the energy conserved? Why?

c) Now consider a new situation in which the cylinder, taken to be hollow and of mass M (same radius R), can spin freely as the mass unwinds. The new angle φ measures the position of P (the place where the mass was at rest) with respect to the vertical axis (see Figure P.1.29b). Write down the Lagrangian in terms of the generalized

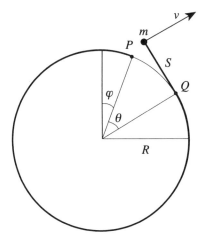

Figure **P.1.29b**

coordinates θ and φ. Identify two conserved quantities and express them as functions of θ and φ.

d) Solve for $\theta(t)$. Which way does the cylinder spin?

1.30   Six Uniform Rods (Stony Brook)

Six equal uniform rods, fastened at their ends by frictionless pivots, form a regular hexagon and lie on a frictionless surface. A blow is given at a right angle to the midpoint of one of them at point P in Figure P.1.30 so that it begins to slide with velocity u. Show that the opposite rod begins to move with velocity $v = u/10$.

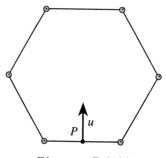

Figure **P.1.30**

1.31   Period as Function of Energy (MIT)

A particle of mass m moves in a one-dimensional potential $U(x) = A|x|^n$, where A is a constant. Find the dependence of the period τ on the energy E in terms of n.

1.32   Rotating Pendulum (Princeton, Moscow Phys-Tech)

The bearing of a rigid pendulum of mass m is forced to rotate uniformly with angular velocity ω (see Figure P.1.32). The angle between the rotation

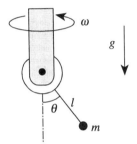

Figure **P.1.32**

axis and the pendulum is called θ. Neglect the inertia of the bearing and of the rod connecting it to the mass. Neglect friction. Include the effects of the uniform force of gravity.

a) Find the differential equation for θ.
b) At what rotation rate ω_c does the stationary point at $\theta = 0$ become unstable?
c) For $\omega > \omega_c$ what is the stable equilibrium value of θ?
d) What is the frequency Ω of small oscillations about this point?

1.33   Flyball Governor (Boston, Princeton, MIT)

Consider the flyball governor for a steam engine shown in Figure P.1.33. Two balls, each of mass m, are attached by means of four hinged arms, each of length l, to sleeves on a vertical rod. The upper sleeve is fastened to the rod; the lower sleeve has mass M and is free to slide up and down the rod as the balls move out from or in toward the rod. The rod-and-ball system rotates with constant angular velocity ω.

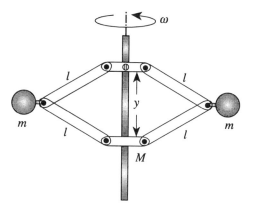

Figure P.1.33

a) Set up the equation of motion, neglecting the weight of the arms and rod. Use as variable the distance y between the sleeves.

b) Show that, for steady rotation of the balls, and $\omega^2 l/g > (1 + M/m)$, the value of the height z of the lower sleeve above its lowest point is

$$z_0 = 2l - \frac{2\,(m + M)\,g}{m\omega^2}$$

c) Show that the angular frequency Ω of small oscillations of z about the steady value z_0 is

$$\Omega = \sqrt{\frac{(m + M)\,g\sin^2\theta_0}{(m + 2M\sin^2\theta_0)\,l\cos\theta_0}}$$

with

$$\theta_0 = \cos^{-1}\left(1 - \frac{z_0}{2l}\right)$$

1.34 Double Pendulum (Stony Brook, Princeton, MIT)

The double pendulum consists of a mass m suspended by a massless string or rod of length l, from which is suspended another such rod and mass (see Figure P.1.34).

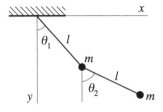

Figure **P.1.34**

a) Write the Lagrangian of the system for $\theta_1, \theta_2 \ll 1$.
b) Derive the equations of motion.
c) Find the eigenfrequencies.

1.35 Triple Pendulum (Princeton)

A triple pendulum consists of masses αm, m, and m attached to a single light string at distances a, $2a$, and $3a$ respectively from its point of suspension (see Figure P.1.35).

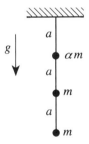

Figure **P.1.35**

a) Determine the value of α such that one of the normal frequencies of this system will equal the frequency of a simple pendulum of length $a/2$ and mass m. You may assume the displacements of the masses from equilibrium are small.
b) Find the mode corresponding to this frequency and sketch it.

1.36 Three Masses and Three Springs on Hoop (Columbia, Stony Brook, MIT)

Three masses, each of mass m, are interconnected by identical massless springs of spring constant k and are placed on a smooth circular hoop as

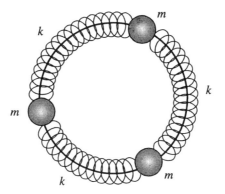

shown in Figure P.1.36. The hoop is fixed in space. Neglect gravity and friction. Determine the natural frequencies of the system, and the shape of the associated modes of vibration.

1.37   Nonlinear Oscillator (Princeton)

a) A nonlinear oscillator has a potential $V(x)$ given by

$$V(x) = \frac{1}{2}kx^2 - \frac{1}{3}m\lambda x^3$$

with λ a small parameter. Find the solution of the equations of motion to first order in λ, assuming $x = 0$ at $t = 0$.

b) Comment on the temperature dependence of the thermal expansion coefficient, if the interaction of the atoms in a solid is described by $V(x)$ from (a).

1.38   Swing (MIT, Moscow Phys-Tech)

A child of mass m on a swing raises her center of mass by a small distance b every time the swing passes the vertical position, and lowers her mass by the same amount at each extremal position. Assuming small oscillations, calculate the work done by the child per period of oscillation. Show that the energy of the swing grows exponentially according to $dE/dt = \alpha E$ and determine the constant α.

1.39   Rotating Door (Boston)

A uniform rectangular door of mass m with sides a and b ($b > a$) and negligible thickness rotates with constant angular velocity ω about a diagonal (see Figure P.1.39). Ignore gravity. Show that the torque $|\mathbf{N}| = \left[m(b^2 - a^2)ab\omega^2\right] / \left[12(a^2 + b^2)\right]$ must be applied to keep the axis of rotation fixed.

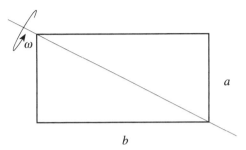

Figure **P.1.39**

1.40 Bug on Globe (Boston)

A toy globe rotates freely without friction with an initial angular velocity ω_0. A bug starting at one pole N travels to the other pole S along a meridian with constant velocity v. The axis of rotation of the globe is held fixed. Let M and R denote the mass and radius of the globe (a solid sphere, moment of inertia $I_0 = 2MR^2/5$), m the mass of the bug, and T the duration of the bug's journey (see Figure P.1.40).

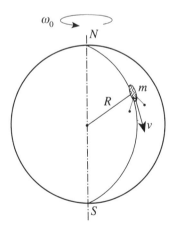

Figure **P.1.40**

Show that, during the time the bug is traveling, the globe rotates through an angle

$$\Delta\theta = \frac{\pi\omega_0 R}{v} \sqrt{2M/(2M + 5m)}$$

A useful integral is

$$\int_0^{2\pi} \frac{dx}{a + b\cos x} = \frac{2\pi}{\sqrt{a^2 - b^2}}, \qquad (a^2 > b^2)$$

1.41   Rolling Coin (Princeton, Stony Brook)

A coin idealized as a uniform disk of radius a with negligible thickness and mass m rolls in a circle. The center of mass of the coin C moves in a circle of radius b and the axis of the coin is tilted at an angle θ with respect to the vertical. Find the angular velocity Ω of the center of mass of the coin (see Figure P.1.41).

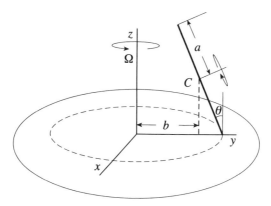

Figure **P.1.41**

1.42   Unstable Top (Stony Brook)

A top of mass M is spinning about a fixed point under gravity, and its axis is vertical ($\theta = 0$, $\dot\theta = 0$), but the angular velocity around its axis ω_3 is insufficient for stability in that position. The Lagrangian for a top is

$$\mathcal{L} = \frac{1}{2}I_1(\dot\theta^2 + \dot\varphi^2 \sin^2\theta) + \frac{1}{2}I_3(\dot\psi + \dot\varphi\cos\theta)^2 - Mgl\cos\theta$$

where θ, φ, ψ are the usual Euler angles, I_1 and I_3 are the moments of

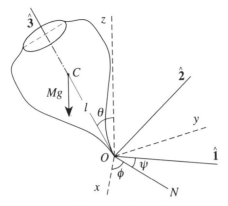

Figure **P.1.42**

inertia about their respective axes, N is the line of nodes, and l is the distance from the point of the top O to the center of mass C (see Figure P.1.42).

a) Derive all the first integrals of the motion and evaluate them in terms of the given initial conditions.

b) Show that the head will descend to an angle θ given by

$$\cos\left(\frac{\theta}{2}\right) = \frac{I_3\omega_3}{2\sqrt{I_1Mgl}}$$

c) Show that the time dependence of this θ is given by the solution of

$$\dot{\theta}^2 = \frac{4Mgl}{I_1}\sin^2\left(\frac{\theta}{2}\right) - \frac{I_3^2\omega_3^2}{I_1^2}\tan^2\left(\frac{\theta}{2}\right)$$

You do not need to solve for $\theta(t)$.

1.43 Pendulum Clock in Noninertial Frame (Maryland)

An off-duty physicist designs a pendulum clock for use on a gravity-free spacecraft. The mechanism is a simple pendulum (mass m at the end of a massless rod of length l) hung from a pivot, about which it can swing

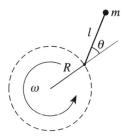

Figure **P.1.43**

in a plane. To provide artificial gravity, the pivot is forced to rotate at a frequency ω in a circle of radius R in the same plane as the pendulum arm (see Figure P.1.43). Show that this succeeds, i.e., that the possible motions $\theta(t)$ of this pendulum are identical to the motions $\theta(t)$ of a simple pendulum in a uniform gravitational field of strength $g = \omega^2 R$, not just for small oscillations, but for any amplitude, and for any length l, even $l > R$.

1.44   Beer Can (Princeton, Moscow Phys-Tech)

A space station is in a circular orbit about the Earth at a radius r_0. An astronaut on a space walk happens to be a distance ϵ on the far side of the station on the line joining the station to the center of the Earth. With practice, the astronaut can throw a beer can so that it appears to orbit

Figure **P.1.44**

the space station in the plane of the motion of the space station about the Earth according to an observer on the station (see Figure P.1.44). You may disregard the gravitational attraction between the beer can and the space station.

a) In what direction and with what velocity relative to the station should the beer can be thrown?
b) What is the period, size and shape of the beer can's orbit, relative to the space station?

1.45 Space Habitat Baseball (Princeton)

On Earth a baseball player can hit a ball 120 m by giving it an initial angle of 45° to the horizontal. Take the acceleration due to gravity as $g = 10$ m/s^2. Suppose the batter repeats this exercise in a space 'habitat' that has the form of a circular cylinder of radius $R = 10$ km and has an angular velocity about the axis of the cylinder sufficient to give an apparent gravity of g at radius R. The batter stands on the inner surface of the habitat (at radius R) and hits the ball in the same way as on Earth (i.e., at 45° to the surface), in a plane perpendicular to the axis of the cylinder (see Figure P.1.45). What is the furthest distance the batter can hit the ball, as measured along the surface of the habitat?

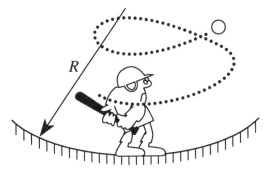

Figure **P.1.45**

1.46 Vibrating String with Mass (Stony Brook)

A thin uniform string of length L and linear density ρ is stretched between two firm supports. The tension in the string is T (see Figure P.1.46).

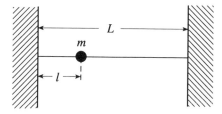

Figure **P.1.46**

a) Derive from first principles the wave equation for small transverse disturbances in the string.
b) Determine the set of possible solutions for the given boundary conditions and state the allowed frequencies.
c) A small mass m is placed a distance l from one end of the string. Determine the first-order correction to the frequencies of the modes found in (b).

1.47 Shallow Water Waves (Princeton (a,b))

Water waves travel on the surface of a large lake of depth d. The lake has a perfectly smooth bottom and the waves are propagating purely in the $+z$ direction (The wave fronts are straight lines parallel to the x axis. See Figure P.1.47).

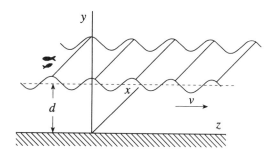

Figure **P.1.47**

a) Find an expression for the velocity of the water $v(y, z, t)$
b) Find the corresponding dispersion relation. You may assume that the flow of the water is irrotational ($\nabla \times \mathbf{v} = 0$), that the amplitude of the waves is small (in practice, this means that $v^2 \ll gh$, where h is the

height of the waves), that surface tension effects are not important, and that water is incompressible.

c) Find the group velocity of the wavefront and consider two limiting cases $\lambda \gg d$, $\lambda \ll d$.

1.48 Suspension Bridge (Stony Brook)

A flexible massless cable in a suspension bridge is subject to uniform loading along the x-axis. The weight of the load per unit length of the cable is w, and the tension in the cable at the center of the bridge (at $x = 0$) is T_0 (see Figure P.1.48).

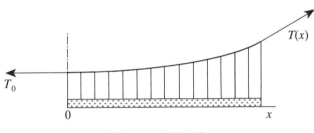

Figure **P.1.48**

a) Find the shape of the cable at equilibrium.
b) What is the tension $T(x)$ in the cable at position x at equilibrium?

1.49 Catenary (Stony Brook, MIT)

A flexible cord of uniform density ρ and fixed length l is suspended from two points of equal height (see Figure P.1.49). The gravitational acceleration is taken to be a constant g in the negative z direction.

a) Write the expressions for the potential energy U and the length l for a given curve $z = z(x)$.
b) Formulate the Euler–Lagrange equations for the curve with minimal potential energy, subject to the condition of fixed length.
c) Show that the solution of the previous equation is given by $z = A \cosh(x/A) + B$, where A and B are constants. Calculate U and l for this solution.

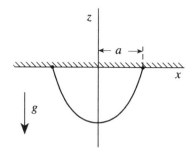

PROBLEMS

Figure **P.1.49**

Formulae:

$$\cosh\xi = \frac{e^\xi + e^{-\xi}}{2} \qquad \sinh\xi = \frac{e^\xi - e^{-\xi}}{2}$$

$$\cosh^2\xi - \sinh^2\xi = 1 \qquad \cosh 2\xi = 2\cosh^2\xi - 1$$

1.50   Rotating Hollow Hoop (Boston)

A thin hollow cylindrical pipe is bent to form a hollow circular ring of mass m and radius R. The ring is attached by means of massless spokes to a vertical axis, around which it can rotate without friction in a horizontal plane. Inside the ring, a point mass P of mass m is free to move without friction, but is connected to a point H of the ring by a massless spring which exerts a force $k\Delta s$, where Δs is the length of the arc HP (see Figure P.1.50). Take as variables the angles θ and φ of CH and CP with the x axis.

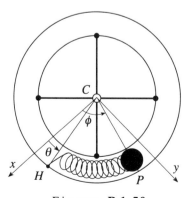

Figure **P.1.50**

a) Write the Lagrangian and the Hamiltonian, and rewrite them in terms of the variables

$$\xi = (\varphi + \theta)/\sqrt{2} \qquad \eta = (\varphi - \theta)/\sqrt{2}$$

b) Find an integral of motion other than the energy, and show that its Poisson bracket with $\mathcal{H}$ is zero.

c) Integrate the equations of motion with these initial conditions at $t = 0$:

$$\theta = -\frac{\pi}{4}, \quad \varphi = +\frac{\pi}{4}, \quad \dot{\theta} = \dot{\varphi} = 0$$

1.51 Particle in Magnetic Field (Stony Brook)

a) Give a relationship between Hamilton's equations under a canonical transformation. Verify that the transformation

$$X = \frac{1}{\sqrt{m\omega}}(\sqrt{2P_1}\sin Q_1 + P_2), \qquad Y = \frac{1}{\sqrt{m\omega}}(\sqrt{2P_1}\cos Q_1 + Q_2)$$

$$P_x = \frac{1}{2}\sqrt{m\omega}(\sqrt{2P_1}\cos Q_1 - Q_2), \qquad P_y = \frac{1}{2}\sqrt{m\omega}(-\sqrt{2P_1}\sin Q_1 + P_2)$$

is canonical.

b) Find Hamilton's equations of motion for a particle moving in a plane in a magnetic field described by the vector potential

$$\mathbf{A} = \left(-\frac{YH}{2}, \frac{XH}{2}, 0\right)$$

in terms of the new variables Q_1, Q_2, P_1, P_2 introduced above, using $\omega = eH/mc$.

1.52 Adiabatic Invariants (Boston (a)) and Dissolving Spring (Princeton, MIT (b))

a) (Adiabatic Invariants) Consider a system with canonical variables

$$p_1, \ldots, p_N, \quad q_1, \ldots, q_N$$

At the time $t = 0$ let C_0 be an arbitrary closed path in phase space and

$$I = \frac{1}{2\pi}\oint_{C_0}\sum_{i=1}^{N} p_i \, dq_i$$

Assume that the point p, q of C_0 moves in phase space according to Hamilton's equations. At a later time the curve C_0 will have become another closed curve C_t. Show that

$$\frac{dI}{dt} = \frac{d}{dt} \frac{1}{2\pi} \oint_{C_t} \sum_{i=1}^{N} p_i \, dq_i = 0$$

and, for a harmonic oscillator with Hamiltonian $\mathcal{H} = (p^2/2m) + (m\omega^2 q^2/2)$, show that

$$I = \frac{E}{\omega}$$

along a closed curve $\mathcal{H}(p, q) = E$.

b) (Dissolving Spring) A mass m slides on a horizontal frictionless track. It is connected to a spring fastened to a wall. Initially, the amplitude of the oscillations is A_1 and the spring constant of the spring is K_1. The spring constant then decreases adiabatically at a constant rate until the value K_2 is reached. (For instance, assume that the spring is being dissolved in acid.) What is the new amplitude?

Hint: Use the result of (a).

1.53   Superball in Weakening Gravitational Field (Michigan State)

A superball is bouncing vertically up and down. It has a velocity v_0 when it strikes the ground. The acceleration due to gravity is slowly reduced by 10% during a very long period of time. Assuming that the collisions of the ball with the ground are elastic, find the corresponding change in v_0.

Relativity

2.1 Marking Sticks (Stony Brook)

Observer O' is travelling with velocity $v = 0.6c$ in the x direction relative to observer O. Each observer has a meter stick with one end fixed at his origin and the other end fixed at x (or x') = 1 m (see Figure P.2.1). Each stick has a marking device (such as a spring-loaded pin) at the high x (or

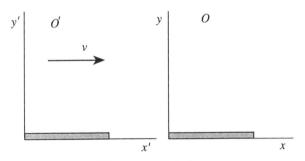

Figure **P.2.1**

x') end, capable of marking the other stick if it overlaps that stick when the marking devices are triggered. The two origins coincide at $t = t' = 0$. Both marking devices are triggered at $t = 0$.

a) According to O, who has the shorter stick? Which stick is marked and where?

b) According to O', who has the shorter stick? Prove by explicit derivation that O' agrees on the result of the marking experiment, including the position of the mark.

2.2   Rockets in Collision (Stony Brook)

A person on Earth observes two rocket ships moving directly toward each other and colliding as shown in Figure P.2.2a. At time $t = 0$ in the Earth

Figure **P.2.2a**

frame, the Earth observer determines that rocket A, travelling to the right at $v_A = 0.8c$, is at point a, and rocket B is at point b, travelling to the left at $v_B = 0.6c$. They are separated by a distance $l = 4.2 \cdot 10^8$ m (see Figure P.2.2b).

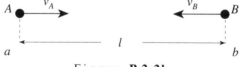

Figure **P.2.2b**

a) In the Earth frame, how much time will pass before the rockets collide?

b) How fast is rocket B approaching in A's frame? How fast is rocket A approaching in B's frame?

c) How much time will elapse in A's frame from the time rocket A passes point a until collision? How much time will elapse in B's frame from the time rocket B passes point b until collision?

2.3   Photon Box (Stony Brook)

An empty box of total mass M with perfectly reflecting walls is at rest in the lab frame. Then electromagnetic standing waves are introduced along the x direction, consisting of N photons, each of frequency ν_0 (see Figure P.2.3).

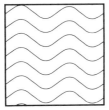

Figure **P.2.3**

a) State what the rest mass of the system (box + photons) will be when the photons are present.
b) Show that this answer can be obtained by considering the momentum and/or energy of the box-plus-photon system in any inertial frame moving along the x axis.

2.4 Cube's Apparent Rotation (Stanford, Moscow Phys-Tech)

A cube with 1-meter edges in its rest frame moves along a straight line at velocity βc. An observer is located in the laboratory frame, and the distance of closest approach is much greater than 1 m. Two faces of the cube are perpendicular to the direction of motion and another two faces are parallel to the plane formed by the trajectory and the observer. The other two faces are approximately perpendicular to the line of sight of the observer (see Figure P.2.4). In this problem, we need to take into account the different travel times for light from different parts of the cube to the observer. This effect causes distortions which make the cube appear to the observer to be rotated. Find the expression for the apparent rotation and indicate the sign of the rotation with respect to the direction of motion of the cube and the line from the cube to the observer.

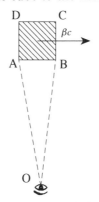

Figure **P.2.4**

2.5   Relativistic Rocket (Rutgers)

A rocket having initially a total mass M_0 ejects its fuel with constant velocity $-u$ ($u > 0$) relative to its instantaneous rest frame. According to Newtonian mechanics, its velocity V, relative to the inertial frame in which it was originally at rest, is related to its mass $M(V)$ by the formula

$$\frac{M}{M_0} = \exp\left(-\frac{V}{u}\right)$$

a) Derive this result.
b) Suppose the velocity of the ejecta is limited only by $0 \le u \le c$, and derive the relativistic analogue of the above equation. Show that it reduces to the Newtonian result at the appropriate limit.

2.6   Rapidity (Moscow Phys-Tech)

a) Consider two successive Lorentz transformations of the three frames of reference K_0, K_1, K_2. K_1 moves parallel to the x axis of K_0 with velocity v, as does K_2 with respect to K_1. Given an object moving in the x direction with velocity v_2 in K_2, derive the formula for the transformation of its velocity from K_2 to K_0.
b) Now consider $n + 1$ frames moving with the same velocity v relative to one another (see Figure P.2.6). Derive the formula for a Lorentz transformation from K_n to K_0, if the velocity of the object in K_n is also v.

Figure   P.2.6

Hint: You may want to use the definition of rapidity or velocity parameter, $\tanh\psi = \beta$, where $\beta = v/c$.

2.7 Charge in Uniform Electric Field (Stony Brook, Maryland, Colorado)

Find the trajectory of a particle of mass m, charge e, in a uniform electric field $\mathbf{E}$, assuming zero velocity parallel to $\mathbf{E}$ at $t = 0$. Sketch the trajectory in the plane of motion.

2.8 Charge in Electric Field and Flashing Satellites (Maryland)

a) Write the relativistic equations of motion for a particle of charge q and mass m in an electromagnetic field. Consider these equations for the special case of motion in the x direction only, in a Lorentz frame that has a constant electric field E pointing in the positive x direction.

b) Show that a particular solution of the equations of motion is given by

$$x = \frac{mc^2}{qE} \cosh\left(\frac{qE\tau}{mc}\right) \qquad t = \frac{mc}{qE} \sinh\left(\frac{qE\tau}{mc}\right) \qquad \text{(P.2.8.1)}$$

and show explicitly that the parameter τ used to describe the world-line of the charge q in equation (P.2.8.1) is the proper time along this worldline.

c) Define the acceleration 4-vector for this motion and show that it has constant magnitude. Draw a space-time (x, ct) diagram showing the worldline (P.2.8.1) and the direction of the acceleration vector at three typical points on the worldline (with $\tau < 0$, $\tau = 0$, $\tau > 0$).

d) Suppose an observer moves along the worldline (P.2.8.1), starting at $t = 0$ and $x = mc^2/qE$. Also, at $\tau = 0$, she leaves behind a satellite that remains at rest at $x = mc^2/qE$. The satellite emits flashes of light at a rate f that is constant in the satellite's rest frame. Show that only a finite number mfc/qE of flashes ever reach the observer.

e) Some time after $\tau = 0$, the observer, always moving along the world-line (P.2.8.1), decides to retrieve the satellite. Show that she cannot wait longer than $t = 3mc/4qE$ or $\tau = (mc/qE)\sinh^{-1}(3/4)$ to decide to do so.

Hint: To retrieve it at this limiting time, she must "reach down" to the satellite with the speed of light, bring it back at the speed of light, and wait indefinitely long for its return.

2.9   Uniformly Accelerated Motion (Stony Brook)

Determine the relativistic uniformly accelerated motion (i.e., the rectilinear motion) for which the acceleration w_0 in the proper reference frame (at each instant of time) remains constant.

a) Show that the 4-velocity

$$u^\mu = \frac{dx^\mu}{d\tau} = \left(\frac{c}{\sqrt{1 - v^2/c^2}}, \frac{\mathbf{v}}{\sqrt{1 - v^2/c^2}} \right) = (c\gamma, \mathbf{v}\gamma)$$

b) Show that the condition for such a motion is

$$w^\mu\, w_\mu = \text{constant} = -w_0^2$$

where w_0 is the usual three dimensional acceleration.

c) Show that in a fixed frame (b) reduces to

$$\frac{d}{dt} \frac{v}{\sqrt{1 - v^2/c^2}} = w_0$$

d) Show that

$$x = \frac{c^2}{w_0} \left(\sqrt{1 + \frac{w_0^2 t^2}{c^2}} - 1 \right)$$

$$v = \frac{w_0 t}{\sqrt{1 + w_0^2 t^2/c^2}}$$

Do these expressions have the correct classical behavior as $c \to \infty$?

2.10   Compton Scattering (Stony Brook, Michigan State)

In the Compton effect, a γ-ray photon of wavelength λ strikes a free, but initially stationary, electron of mass m. The photon is scattered an angle θ, and its scattered wavelength is $\tilde{\lambda}$. The electron recoils at an angle φ (see Figure P.2.10).

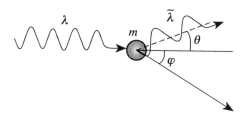

Figure **P.2.10**

a) Write the relativistic equations for momentum and energy conservation.

b) Find an expression for the change $\tilde{\lambda} - \lambda$ in the photon wavelength for the special case $\theta = \pi/2$.

2.11   Mössbauer Effect (Moscow Phys-Tech, MIT, Colorado)

An atom in its ground state has mass m. It is initially at rest in an excited state of excitation energy $\Delta\mathcal{E}$. It then makes a transition to the ground state by emitting one photon. Find the frequency of the photon, taking into account the relativistic recoil of the atom. Express your answer also in terms of the mass M of the excited atom. Discuss this result for the case of a crystalline lattice (Mössbauer effect).

2.12   Positronium and Relativistic Doppler Effect (Stony Brook)

An electron e^- and a positron, e^+, each of mass m_e, bound with binding energy $\mathcal{E}_b$ in positronium, annihilate into two photons.

a) Calculate the energy, momentum, velocity, and frequency of the photons.

b) The positronium with velocity $\mathbf{v}$ moves away from the observer in the lab and annihilates as shown in Figure P.2.12. Give the frequency of the photon as measured by the observer and calculate its frequency in terms of the frequency in the positronium rest system (Relativistic Doppler Effect).

Figure **P.2.12**

2.13   Transverse Relativistic Doppler Effect (Moscow Phys-Tech)

A qualitative difference between classical mechanics and relativity is the existence of the transverse Doppler effect in relativity (when light propagates perpendicular to its source in the observer's frame). Calculate the frequency of the photon ω' in the observer's frame in terms of its frequency ω in the rest frame.

2.14   Particle Creation (MIT)

Consider a photon of energy $\mathcal{E}_\gamma$ incident on a stationary proton. For sufficiently large $\mathcal{E}_\gamma$, a π meson can be produced in a reaction

$$\gamma + p \rightarrow p + \pi^0$$

What is $\mathcal{E}_{\min}$, the threshold photon energy for this reaction to occur?

2.15   Electron–Electron Collision (Stony Brook)

An electron of total energy 1.40 MeV collides with another electron which is at rest in the laboratory frame. Let the electronic rest energy be 0.51 MeV.

a) What are the total energy and momentum of the system in the laboratory frame of reference (specify momentum in MeV/c units)?

b) Find the velocity of the center of mass in the laboratory frame.

c) Determine the total energy of the pair of particles in the center of mass frame of reference (CMF).

d) The target electron scatters at an angle of 45° in the CMF. What will be the direction of scatter of the projectile in the CMF? What will be the energy and momentum of the target electron after scatter in the CMF?

e) What, in the laboratory frame, will be the target electron's momentum components perpendicular and parallel to the direction of the incoming particle after the collision?

2.16   Inverse Compton Scattering (MIT, Maryland)

The HERA electron storage ring at Hamburg, Germany contains circulating electrons with an energy of 27 GeV. Photons of wavelength 514 nm from an argon-ion laser are directed so as to collide head-on with the stored electrons. Calculate the maximum scattered photon energy.

2.17   Proton–Proton Collision (MIT)

A proton with $\gamma = 1/\sqrt{1 - (v^2/c^2)}$ collides elastically with a proton at rest. If the two protons rebound with equal energies, what is the angle θ between them?

2.18 Pion Creation and Neutron Decay (Stony Brook)

a) Calculate the threshold energy in MeV for the creation of charged pions by photons incident on stationary protons,

$$\gamma + p \rightarrow n + \pi^+$$

b) Neutrons β-decay via

$$n \rightarrow p + e^- + \bar{\nu}_e$$

If the neutron is at rest, what is the maximum possible velocity for the electron in units of c? (Note that $m_p = 938.27 \, \text{MeV}/c^2$, $m_n = 939.57 \, \text{MeV}/c^2$, $m_{\pi^+} = 139.57 \, \text{MeV}/c^2$, and $m_e = 0.51 \, \text{MeV}/c^2$.)

2.19 Elastic Collision and Rotation Angle (MIT)

Consider an elastic collision (namely a collision where the particles involved do not change their internal state) of an incident particle of mass m_1, momentum $\mathbf{p}_1$, and energy $\mathcal{E}_1$ (see notation below), with a particle of mass m_2 at rest. Let the final energies be $\tilde{\mathcal{E}}_1$ and $\tilde{\mathcal{E}}_2$ and the final momenta be $\tilde{\mathbf{p}}_1$ and $\tilde{\mathbf{p}}_2$ (all of this in the laboratory frame).

a) In the center of mass frame (CMF), denote by $\mathbf{p}_{10}$ and $\mathbf{p}_{20}$ the incoming momenta of the two particles and by $\mathcal{E}_{10}, \mathcal{E}_{20}$ their energies. One has

$$\mathbf{p}_{10} = -\mathbf{p}_{20} = \mathbf{p}_0, \qquad |\mathbf{p}_0| \equiv p_0 \qquad \text{(P.2.19.1)}$$

where $\mathbf{p}_0$ is a vector. In the center of mass frame the collision rotates the direction of the momenta. Let the outgoing momenta and energies be $\tilde{\mathbf{p}}_{10}, \tilde{\mathbf{p}}_{20}, \tilde{\mathcal{E}}_{10}, \tilde{\mathcal{E}}_{20}$, and the rotation angle be χ. From conservation of energy and momentum, what can you tell about $\tilde{\mathbf{p}}_{10}, \tilde{\mathbf{p}}_{20}$? About $\tilde{\mathcal{E}}_{10}$ and $\tilde{\mathcal{E}}_{20}$?

b) From the energy and momentum conservation laws,

$$p_{1\mu} + p_{2\mu} = \tilde{p}_{1\mu} + \tilde{p}_{2\mu} \qquad \text{(P.2.19.2)}$$

show that

$$p_1 \cdot p_2 - p_1 \cdot \tilde{p}_1 - p_2 \cdot \tilde{p}_1 + m_1^2 = 0 \qquad \text{(P.2.19.3)}$$

c) Evaluate the first and third terms of the left-hand side of equation (P.2.19.3) in the laboratory frame. Evaluate the second term in the CMF in terms of p_0 and χ (and masses). Now use equation (P.2.19.3) to find an expression for $(\tilde{\mathcal{E}}_1 - \mathcal{E}_1)$ in terms of p_0 and χ (and masses).

d) Find an expression for p_0 in terms of $\mathcal{E}_1$ (and masses) by evaluating $p_1 \cdot p_2$ both in the laboratory and CMF.

e) Give $\tilde{\mathcal{E}}_1$ and $\tilde{\mathcal{E}}_2$ in terms of $\mathcal{E}_1, \mathcal{E}_2$, and χ.

f) Consider the case for maximal energy transfer. What is the value of χ? For this case find the ratio of the final kinetic energy to the incident kinetic energy for the incoming particle (in the laboratory frame).

Notation: Here we use

$$c = 1$$

$$p \cdot p = p_\mu p^\mu = \mathcal{E}_0^2 - \mathbf{p}^2 = m^2$$

$$p_1 \cdot p_2 = p_{1\mu} p_2^\mu = p_1^\mu p_{2\mu}$$

$$\text{kinetic energy} = \mathcal{E} - m$$

Electrodynamics

3.1 Charge Distribution (Wisconsin-Madison)

An electric charge distribution produces an electric field

$$\mathbf{E} = c\left(1 - e^{-\alpha r}\right)\frac{\hat{\mathbf{r}}}{r^2}$$

where c and α are constants. Find the net charge within the radius $r = 1/\alpha$.

3.2 Electrostatic Forces and Scaling (Moscow Phys-Tech)

a) Consider two solid dielectric spheres of radius a, separated by a distance R ($R \gg a$). One of the spheres has a charge q, and the other is neutral (see Figure P.3.2a). We scale up the linear dimensions of the system by a factor of two. How much charge should reside on the first sphere now so that the force between the spheres remains the same?

b) Now consider a conducting ring made of thin wire, where d is the diameter of the wire and D is the diameter of the ring (again, $D \gg d$). A charge Q placed on the ring is just sufficient to cause the ring to

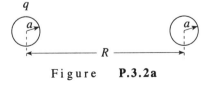

Figure   P.3.2a

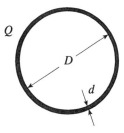

Figure   P.3.2b

break apart due to electrostatic repulsion (see Figure P.3.2b). As in
(a), the linear dimensions of the system are multiplied by two. At
what charge will the new ring break?

3.3   Dipole Energy (MIT, Moscow Phys-Tech)

An electric dipole of moment **p** is placed at a height h above a perfectly
conducting plane and makes an angle of θ with respect to the normal to
this plane (see Figure P.3.3).

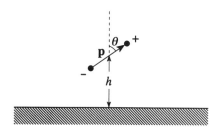

Figure   P.3.3

a) Indicate the position and orientation of the image dipole and the
   direction of the force felt by the dipole.
b) Calculate the work required to remove the dipole to infinity.

3.4   Charged Conducting Sphere in Constant Electric Field (Stony Brook, MIT)

A conducting sphere of radius a on whose surface resides a total charge Q is
placed in a uniform electric field $\mathbf{E}_0$ (see Figure P.3.4). Find the potential

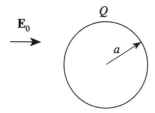

Figure   **P.3.4**

at all points in space exterior to the sphere. What is the surface charge density?

3.5   Charge and Conducting Sphere I (MIT)

A point charge e is placed at a distance R from the center of a metallic sphere of radius a, with $R > a$ (see Figure P.3.5). The sphere is insulated and is electrically neutral.

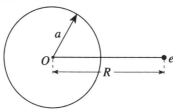

Figure   **P.3.5**

a) Find the electrostatic potential on the surface of the sphere.
b) Find the force acting on the charge.

3.6   Charge and Conducting Sphere II (Boston)

A charge e is placed at a distance R from the center of a grounded conducting sphere of radius $a < R$ (see Figure P.3.6). Show that the force acting

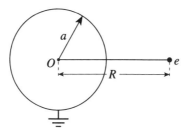

Figure   **P.3.6**

on the charge is proportional to $1/R^3$ when $R \gg a$ and $1/(R-a)^2 = 1/\delta^2$ when $R = a + \delta$, $0 < \delta \ll a$.

3.7   Conducting Cylinder and Line Charge (Stony Brook, Michigan State)

The axis of a long, thin-walled, conducting, uncharged cylindrical shell of radius a is oriented along the z-axis as shown in Figure P.3.7. A long, thin wire carrying a uniform linear charge density $+\lambda$ runs parallel to the cylinder, at a distance R from the center. Use the method of images to find the electric potential in the x–y plane.

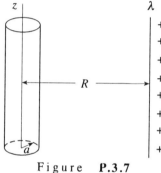

Figure   P.3.7

a) State the conditions that have to be met by the image charge. Find the potential on the surface of the shell relative to infinity.

b) Find the potential at any point ρ, φ in the x–y plane outside the cylinder.

Hint: It is possible to find an image charge such that the potential at infinity in the x–y plane is zero.

3.8   Spherical Void in Dielectric (Princeton)

Suppose there is a spherical void of radius R in an otherwise homogeneous material of dielectric constant ε (see Figure P.3.8). At the center of the void is a point dipole **p**. Solve for the electric field everywhere.

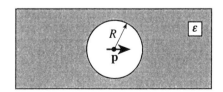

Figure   P.3.8

3.9   Charge and Dielectric (Boston)

A charge e is situated at the point $x = h > 0$, $y = z = 0$ outside a homogeneous dielectric which fills the region $x < 0$ (see Figure P.3.9).

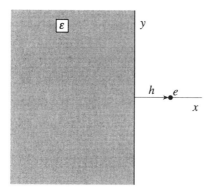

F i g u r e   P.3.9

a) Write the electric fields $\mathbf{E}(0^+, y, z)$ and $\mathbf{E}(0^-, y, z)$ just outside and just inside the dielectric in terms of the charge e and surface charge density σ_b of bound charges on the surface of the dielectric.

b) Express σ_b in terms of $\mathbf{E}(0^-, y, z)$. Denote by ε the dielectric constant of the dielectric.

c) By using the equations obtained in (a) and (b), show that

$$\sigma_b = -\frac{1}{2\pi}\frac{\varepsilon - 1}{\varepsilon + 1}\frac{eh}{(h^2 + y^2 + z^2)^{3/2}}$$

d) Calculate the electric field $\mathbf{E}'$ due to σ_b at the position $(h, 0, 0)$ of the charge e. Show that it can be interpreted as the field of an image charge e' situated at the point $(-h, 0, 0)$.

e) Show that the charge e experiences the force

$$\mathbf{F} = -\frac{\varepsilon - 1}{\varepsilon + 1}\frac{e^2}{4h^2}\hat{\mathbf{x}}$$

3.10   Dielectric Cylinder in Uniform Electric Field (Princeton)

An infinitely long circular cylinder of radius a, dielectric constant ε, is placed with its axis along the z-axis, and in an electric field which would be uniform in the absence of the cylinder, $\mathbf{E} = E_0\hat{\mathbf{x}}$ (see Figure P.3.10). Find

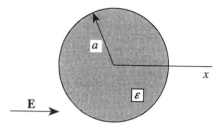

Figure   **P.3.10**

the electric field at points outside and inside the cylinder and the bound surface charge density.

3.11   Powder of Dielectric Spheres (Stony Brook)

A powder composed of small spherical particles (with $\varepsilon = 4$ and of radius $R = 100$ nm) is dispersed in vacuum with a concentration of $n = 10^{12}$ particles per cm^3. Find the effective dielectric constant ε' of this medium. Explain why the apparent answer,

$$\varepsilon' = 1 + (\varepsilon - 1)nV$$

(where $V = 4\pi R^3/3$ is the volume of one particle) is wrong.

Hint: Make use of the fact that $nR^3 \ll 1$. Exploit the spherical symmetry of the particles.

3.12   Concentric Spherical Capacitor (Stony Brook)

Consider two concentric metal spheres of finite thickness in a vacuum. The inner sphere has radii $a_1 < a_2$. The outer sphere has $b_1 < b_2$ (see Figure P.3.12).

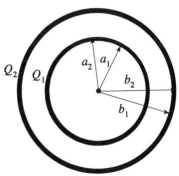

Figure   **P.3.12**

a) A charge Q_1 is put on the inner sphere and a charge Q_2 on the outer sphere. Find the charge density on each of the four surfaces. If $Q_2 = -Q_1$, what is the mutual capacitance of the system?

b) If the space between the spheres is filled with insulating material of dielectric constant ε_1, what are the surface charge densities and polarization surface charge densities for arbitrary Q_1 and Q_2 and the mutual capacitance for $Q_2 = -Q_1$?

3.13   Not-so-concentric Spherical Capacitor (Michigan Tech)

An insulated metal sphere of radius a with total charge q is placed inside a hollow grounded metal sphere of radius b. The center of the inner sphere is slightly displaced from the center of the outer sphere so that the distance between the two centers is δ (see Figure P.3.13).

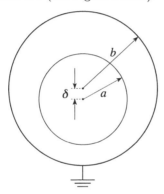

Figure **P.3.13**

a) Use the boundary conditions to determine the potential between the spheres in the case $\delta = 0$.

b) Find the charge distribution of the inner sphere and the force acting on it.

Hint: Show that $R(\theta) \approx b + \delta \cos \theta$, where R is the distance from the center of the inner sphere to the surface of the outer sphere, and write down an expansion for the potential between the spheres using spherical harmonics to first order in δ.

3.14   Parallel Plate Capacitor with Solid Dielectric (Stony Brook, Michigan Tech, Michigan)

Two square metal plates of side L are separated by a distance $d \ll L$. A dielectric slab of size $L \times L \times d$ just slides between the plates. It is inserted a distance x (parallel to one side of the squares) and held there (see Figure

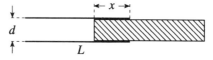

P.3.14). The metal plates are then charged to a potential difference V and disconnected from the voltage source.

a) Find the force exerted electrically on the slab. Be careful and explicit about its direction.
b) How does the situation change if the battery is left connected?

3.15   Parallel Plate Capacitor in Dielectric Bath (MIT)

A parallel plate capacitor with square plates of side L and plate separation d is charged to a potential V and disconnected from the battery. It is then vertically inserted into a large reservoir of dielectric liquid with relative dielectric constant ε and density ρ until the liquid fills half the space between the capacitor plates as shown in Figure P.3.15.

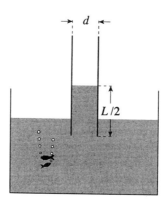

Figure P.3.15

a) What is the capacitance of the system?
b) What is the electric field strength between the capacitor plates?
c) What is the distribution of charge density over the plates?
d) What is the difference in vertical height between the level of liquid within the capacitor plates and that in the external reservoir?

3.16 Not-so-parallel Plate Capacitor (Princeton (a), Rutgers (b))

a) A capacitor is formed by two rectangular conducting plates having edges L_1 and L_2. The plates are not parallel. One pair of edges of length L_1 is separated by a distance d_1 everywhere, and the other pair of edges of length L_1 is separated by d_2 everywhere; $d_2 > d_1$

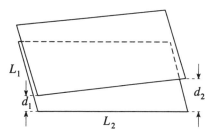

Figure **P.3.16**

(see Figure P.3.16). Neglecting edge effects, when a voltage difference V is placed across the two conductors, find the potential everywhere between the plates.

b) Determine the capacitance.

3.17 Cylindrical Capacitor in Dielectric Bath (Boston, Maryland)

The electrostatic field energy U_e of a capacitor can be expressed as a function of a parameter x (e.g., the plate separation) and the fixed plate charge (no charging battery present) or as a function of x and of the electromotive force V_b of a battery to which the plates are connected.

a) Show that the generalized force F_{ex} corresponding to the parameter x is given by

$$F_{ex} = -\frac{dU_e(x, Q)}{dx} = +\frac{dU_e(x, V_b)}{dx}$$

b) Verify these formulae for the case of a parallel plate capacitor.

c) A cylindrical capacitor is lowered vertically into a reservoir of liquid dielectric of mass density ρ. If a voltage V is applied between the inner cylinder (radius a) and the outer shell (radius b), the liquid

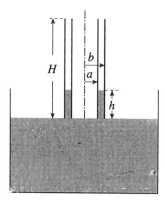

Figure P.3.17

rises to a height h between them (see Figure P.3.17). Show that

$$h = \frac{(\varepsilon - 1)V^2}{4\pi \rho g(b^2 - a^2) \ln(b/a)}$$

3.18 Iterated Capacitance (Stony Brook)

a) Given two point charges of opposite sign and unequal in magnitude, show that the (nontrivial!) surface with $V = V(\infty)$ is a sphere. Find its radius and center. This is the basis of the "method of images" for problems involving equipotential spheres.

b) Describe concisely but clearly an iterative method to find the capacitance of two conducting spheres of radius a whose centers are $4a$ apart.

c) Calculate the capacitance to within 5%.

3.19 Resistance vs. Capacitance (Boston, Rutgers (a))

a) Consider two conductors of some shape. Use them in two alternative ways, as a capacitor and as a resistor as shown in (a) and (b) of Figure P.3.19, respectively. In case (a), the space between the conductors is filled with a homogeneous material of permittivity ε, while in case (b), it is filled with a homogeneous material of finite conductivity σ. By considering separately these two cases, prove the relation

$$RC = \frac{\varepsilon}{4\pi\sigma}$$

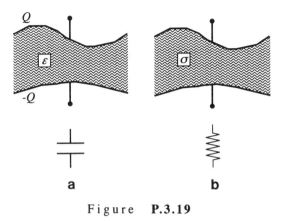

Figure   **P.3.19**

between the capacitance C in case (a) and the resistance R in case (b). If you cannot give a general proof, try at least some special example, e.g., parallel plates.

b) Two conducting spheres have their centers a distance c apart. Their respective radii are a and b. Show that when $c \gg a, b$ the capacitance of this system will be given approximately by

$$C \approx \left(\frac{1}{a} + \frac{1}{b} - \frac{2}{c} \right)^{-1}$$

c) Two small, spherical, perfectly conducting electrodes of radii a and b are embedded in an infinite medium of conductivity σ. Their centers are separated by a distance $c \gg a, b$. Find the resistance between them without using (a) and (b).

Hint: If two electrodes at potentials V_1 and V_2 are embedded in a medium of finite conductivity, the currents I_1 and I_2 leaving each of them are related to the potentials by the formulae $V_1 = R_{11}I_1 + R_{12}I_2$, $V_2 = R_{21}I_1 + R_{22}I_2$. Determine the coefficients R_{ij} by considering cases with $I_2 = 0$ and $I_1 = 0$.

d) Check the results of (b) and (c) by using (a).

3.20   Charge Distribution in Inhomogeneous Medium (Boston)

A stationary current distribution is established in a medium that is isotropic but not necessarily homogeneous. Show that the medium will in general acquire a volume distribution of charge whose density is (in Gaussian units)

$$\rho = -\frac{1}{4\pi\sigma}(\sigma\nabla\varepsilon - \varepsilon\nabla\sigma)\cdot\nabla\phi$$

where σ and ε are the conductivity and the dielectric permittivity of the medium and ϕ is the potential.

3.21   Green's Reciprocation Theorem (Stony Brook)

a) Prove Green's reciprocation theorem: If ϕ is the potential due to a volume charge density ρ within a volume V and a surface charge density σ on the conducting surface S bounding the volume V, while ϕ' is the potential due to another charge distribution ρ' and σ', then

$$\int_V \rho\phi'\, d^3x + \int_S \sigma\phi'\, dS = \int_V \rho'\phi\, d^3x + \int_S \sigma'\phi\, dS$$

b) A point charge q is placed between two infinite grounded parallel conducting plates. If z_0 is the distance between q and the lower plate, find the total charge induced on the upper plate in terms of q, z_0, and l, where l is the distance between the plates (see Figure P.3.21). Show your method clearly.

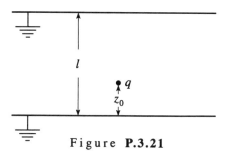

Figure P.3.21

3.22   Coaxial Cable and Surface Charge (Princeton)

A very long coaxial cable consists of an inner cylinder of radius a and isotropic conductivity σ and a concentric outer cylinder of radius b. The outer shell has infinite conductivity. The space between the cylinders is empty. A uniform, constant current density $\mathbf{J}$, directed along the axial coordinate z, is maintained in the inner cylinder. Return current flows

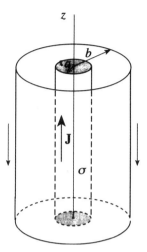

Figure **P.3.22**

uniformly in the outer shell (see Figure P.3.22). Compute the surface charge density on the inner cylinder as a function of the axial coordinate z, with the origin $z = 0$ chosen to be the plane halfway between the two ends of the cable.

3.23   Potential of Charged Rod (Stony Brook)

A thin nonconducting rod of length L carries a uniformly distributed charge Q and is oriented as shown in Figure P.3.23.

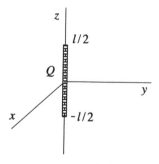

Figure **P.3.23**

a) Find the potential ϕ due to the charged rod for any point on the z-axis with $z > L/2$.

b) Find $\phi(r, \theta, \varphi)$ for all $|\mathbf{r}| > L/2$ where r, θ, φ are the usual spherical coordinates.

Hint: The general solution to Laplace's equation in spherical coordinates is

$$\phi(r,\theta,\varphi) = \sum_{l=0}^{\infty} \sum_{m=-l}^{l} \left[A_l r^l + \frac{B_l}{r^{l+1}} \right] Y_{lm}(\theta,\varphi) \qquad \text{(P.3.23.1)}$$

3.24 Principle of Conformal Mapping (Boston)

a) Show that the real part $U(x,y)$ and the imaginary part $V(x,y)$ of a differentiable function $W(z)$ of $z = x + iy$ obey Laplace's equation.
b) If $U(x,y)$ and $V(x,y)$ above are the potentials of two fields $\mathbf{F}$ and $\mathbf{G}$ in two dimensions, show that at each point (x,y), the fields $\mathbf{F}$ and $\mathbf{G}$ are orthogonal.
c) Consider the function $W(z) = A \ln z$, where A is a real constant. Find the fields $\mathbf{F}$ and $\mathbf{G}$ and mention physical (Electrodynamics) problems in which they might occur.

3.25 Potential above Half Planes (Princeton)

An infinite conducting plane (the x–z plane in Figure P.3.25) is divided by the line $z = 0$. For $x > 0$, the potential in the plane is $+V_0$, while for $x < 0$. the potential is $-V_0$. Evaluate the potential everywhere.

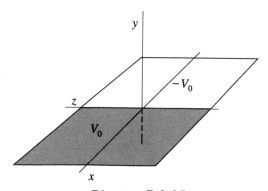

Figure **P.3.25**

3.26 Potential of Halved Cylinder (Boston, Princeton, Chicago)

Consider an infinitely long conducting cylinder of radius a with its axis coinciding with the z-axis. One half of the cylinder (cut the long way) $(y > 0)$ is kept at a constant potential V_0, while the other half $(y < 0)$ is

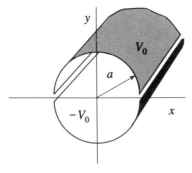

Figure **P.3.26**

kept at a constant potential $-V_0$ (see Figure P.3.26). Find the potential for all points inside the cylinder and the field **E** along the z-axis.

3.27 Resistance of a Washer (MIT)

A washer is made of a dielectric of resistivity ρ. It has a square cross section of length a on a side, and its outer radius is $2a$. A small slit is made on one side and wires of negligible resistance are connected to the faces exposed by the slit (see Figure P.3.27). If the wires were connected into a circuit, what would be the lumped resistance due to the washer?

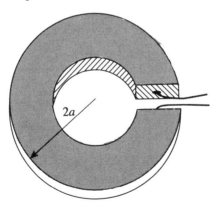

Figure **P.3.27**

3.28 Spherical Resistor (Michigan State)

A and B are opposite ends of a diameter AOB of a very thin spherical shell of radius a and thickness t. Current enters and leaves by two small

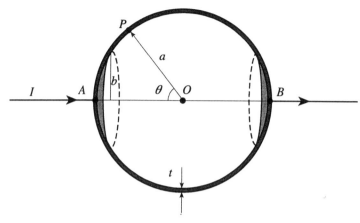

Figure **P.3.28**

circular electrodes of radius b whose centers are at A and B (see Figure
P.3.28). If I is the total current and P is a point on the shell such that the
angle $POA = \theta$, show that the magnitude of the current density vector at P
is proportional to $(2\pi at \sin\theta)^{-1}$. Hence find the resistance of the conductor.

You may find this integral useful:

$$\int \frac{dx}{\sin x} = -\frac{1}{2}\ln\left[\frac{1+\cos x}{1-\cos x}\right]$$

3.29   Infinite Resistor Ladder (Moscow Phys-Tech)

Consider the ladder of resistors, each of resistance r, shown in Figure P.3.29.
What is the resistance seen between terminals A and C?

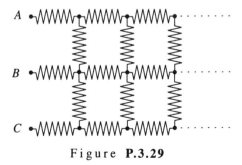

Figure **P.3.29**

3.30 Semi-infinite Plate (Moscow Phys-Tech)

Consider a thin semi-infinite plate of negligible thickness made of an isotropic conductive material. A voltage $V_0 = 1$ V is applied across points A and B of the plate (see Figure P.3.30). At a distance $d = 1$ cm from the end a voltage of 0.1 V is measured between points C and D. Find the voltage difference between two analogous points an arbitrary distance x from the end of the plate.

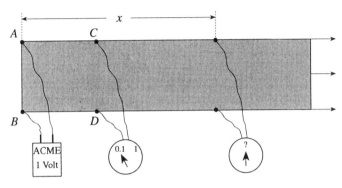

Figure **P.3.30**

3.31 Magnetic Field in Center of Cube (Moscow Phys-Tech)

The current I flowing along the edges of one face of a cube (see Figure P.3.31a) produces a magnetic field in the center of the cube of magnitude B_0. Consider another cube where the current I flows along a path shown in Figure P.3.31b. What magnetic field will now exist at the center of the cube?

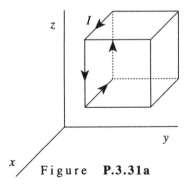

Figure **P.3.31a**

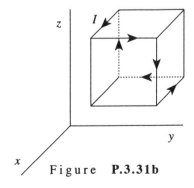

Figure **P.3.31b**

3.32   Magnetic Dipole and Permeable Medium (Princeton)

A point magnetic dipole **m** in vacuum (medium 1) is pointing toward the plane surface of a medium with permeability μ (medium 2). The distance between the dipole and surface is d (see Figure P.3.32).

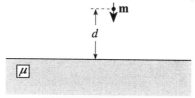

Figure P.3.32

a) Solve for the magnetic field **B** within the medium.
b) What is the force acting on the dipole?

3.33   Magnetic Shielding (Princeton)

A spherical shell of high permeability μ is placed in a uniform magnetic field.

a) Compute the attenuation (shielding) produced by the sphere in terms of μ and the inner and outer radii a and b, respectively, of the shell.
b) Take the limit at $\mu \gg 1$ and estimate the shielding for $\mu = 10^5$, $a = 0.5$ m, $b = 0.55$ m.

3.34   Electromotive Force in Spiral (Moscow Phys-Tech)

A flat metal spiral (with a constant distance h between coils) and a total number of coils N is placed in a uniform magnetic field $B = B_0 \cos \omega t$ perpendicular to the plane of the spiral (see Figure P.3.34). Evaluate the total electromotive force induced in the spiral (between points A and C). Assume $N \gg 1$.

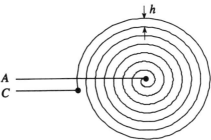

Figure P.3.34

3.35   Sliding Copper Rod (Stony Brook, Moscow Phys-Tech)

A copper rod slides on frictionless rails in the presence of a constant magnetic field $\mathbf{B} = B_0\hat{\mathbf{z}}$. At $t = 0$, the rod is moving in the y direction with velocity v_0 (see Figure P.3.35).

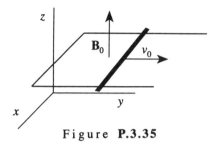

Figure P.3.35

a) What is the subsequent velocity of the rod if σ is its conductivity and ρ_m the mass density of copper.

b) For copper, $\sigma = 5 \times 10^{17}\text{s}^{-1}$ and $\rho_m = 8.9\,\text{g/cm}^3$. If $B_0 = 1$ gauss, estimate the time it takes the rod to stop.

c) Show that the rate of decrease of the kinetic energy of the rod per unit volume is equal to the ohmic heating rate per unit volume.

3.36   Loop in Magnetic Field (Moscow Phys-Tech, MIT)

A conducting circular loop made of wire of diameter d, resistivity ρ, and mass density ρ_m is falling from a great height h in a magnetic field with a component $B_z = B_0(1 + \kappa z)$, where κ is some constant. The loop of diameter D is always parallel to the x–y plane. Disregarding air resistance, find the terminal velocity of the loop (see Figure P.3.36).

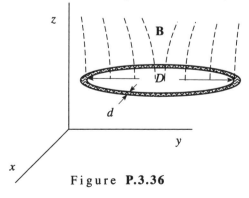

Figure P.3.36

3.37 Conducting Sphere in Constant Magnetic Field (Boston)

A perfectly conducting sphere of radius R moves with constant velocity $\mathbf{v} = v\hat{\mathbf{x}}$ ($v \ll c$) through a uniform magnetic field $\mathbf{B} = B_0\hat{\mathbf{y}}$ (see Figure P.3.37). Find the surface charge density induced on the sphere to lowest order in v/c.

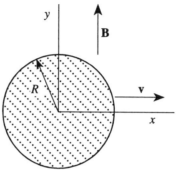

Figure   P.3.37

3.38 Mutual Inductance of Line and Circle (Michigan)

A circular wire of radius a is insulated from an infinitely long straight wire in a tangential direction (see Figure P.3.38). Find the mutual inductance.

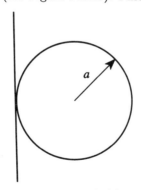

Figure P.3.38

3.39 Faraday's Homopolar Generator (Stony Brook, Michigan)

Consider a perfectly conducting disk of radius r_0 in a constant magnetic field B perpendicular to the plane of the disk. Sliding contacts are provided

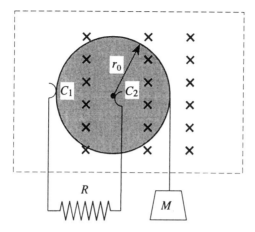

Figure **P.3.39**

at the edge of the disk (C_1) and at its axle (C_2) (see Figure P.3.39). This system is Faraday's "homopolar generator." When turned at constant angular velocity, it provides a large direct current with no ripple. A torque is produced by a mass M hung on a long string wrapped around the perimeter of the disk.

a) Explain how and why a current flows. Give a quantitative expression for the current as a function of angular velocity.

b) Given a long enough string, this system will reach a constant angular velocity ω_f. Find this ω_f and the associated current.

3.40 Current in Wire and Poynting Vector (Stony Brook, MIT)

A long straight wire of radius b carries a current I in response to a voltage V between the ends of the wire.

a) Calculate the Poynting vector **S** for this DC voltage.

b) Obtain the energy flux per unit length at the surface of the wire. Compare this result with the Joule heating of the wire and comment on the physical significance.

3.41 Box and Impulsive Magnetic Field (Boston)

Two opposite walls of a rigid box are uniformly charged with surface charge densities σ and $-\sigma$, respectively. The positively charged wall occupies the region $0 \leq x \leq a$, $0 \leq y \leq b$ of the plane $z = h$, while the negatively charged wall occupies the region $0 \le x \le a$, $0 \le y \le b$ of the $x\text{-}y$ plane. Inside the box, there is a uniform magnetic field $\mathbf{B} = (0, B_0, 0)$. Assume that h is much smaller than both a and b and that the charged walls are nonconducting (see Figure P.3.41).

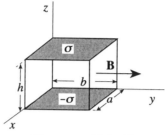

Figure **P.3.41**

a) Estimate the impulse experienced by the box when the magnetic field is switched off.

b) Show that it is equal to the initial momentum of the electromagnetic field.

3.42 Coaxial Cable and Poynting Vector (Rutgers)

The infinitely long coaxial line in Figure P.3.42 carries a steady current I upwards in the inner conductor and a return current I downwards in the outer conductor. Both conductors have a resistance per length (along the axes) λ. The space between the inner and outer conductors is occupied by

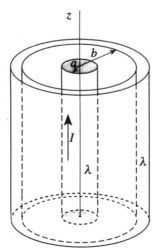

Figure **P.3.42**

a vacuum. The radius of the inner conductor is a, and that of the outer conductor is b. In the following, use the cylindrical coordinates, ρ, φ, z. In these coordinates,

$$\nabla^2 \phi = \frac{1}{\rho} \frac{\partial}{\partial \rho} \left(\rho \frac{\partial \phi}{\partial \rho} \right) + \frac{1}{\rho^2} \frac{\partial^2 \phi}{\partial \varphi^2} + \frac{\partial^2 \phi}{\partial z^2}$$

a) Find the electrostatic potential and the electric field in the region $a < \rho < b$. Assume that $E_\rho(\rho, \varphi, 0) = 0$.
b) Find the magnetic field in the region $a < \rho < b$.
c) Calculate the Poynting vector in the region $a < \rho < b$ and integrate it over the surface of the volume bounded by $\rho = a$, $\rho = b$, and $-l/2 \leq z \leq l/2$. Comment on the physical implications of your result.

3.43 Angular Momentum of Electromagnetic Field (Princeton)

Consider two spherical metal shells of radii a and b (see Figure P.3.43). There is a magnetic dipole of moment $\mathbf{M}$ in the center of the inner sphere. There is a charge $+q$ on the inner sphere and $-q$ on the outer sphere. Find the angular momentum associated with the electromagnetic field of the system.

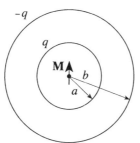

Figure **P.3.43**

3.44 Plane Wave in Dielectric (Stony Brook, Michigan)

A monochromatic plane wave of frequency ω propagates through a nonpermeable ($\mu = 1$) insulating medium with dielectric constant ε_1. The wave is normally incident upon an interface with a similar medium with dielectric constant ε_2 (see Figure P.3.44).

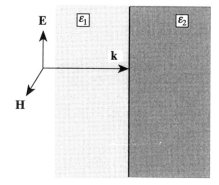

Figure P.3.44

a) Derive the boundary conditions for the electric and magnetic fields at the interface.
b) Find the fraction of incident energy that is transmitted to the second medium.

3.45   X-Ray Mirror (Princeton)

X-rays which strike a metal surface at an angle of incidence to the normal greater than a critical angle θ_0 are totally reflected. Assuming that a metal contains n free electrons per unit volume, calculate θ_0 as a function of the frequency ω of the X-rays. The metal occupies the region $x > 0$. The X-rays are propagating in the x–y plane (the plane of the picture) and their polarization vector is in the z direction, coming out of the picture (see Figure P.3.45).

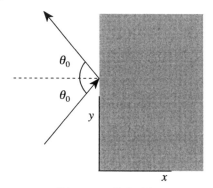

Figure P.3.45

3.46   Plane Wave in Metal (Colorado, MIT)

Suppose that a plane electromagnetic wave of frequency $\omega/2\pi$ and amplitude E_0 is normally incident on the flat surface of a semi-infinite metal of conductivity σ. Assume the frequency is low so that the displacement current inside the metal can be neglected. The magnetic permeability of the metal $\mu = 1$.

a) Using Maxwell's equations, derive expressions for the components of the electric and magnetic fields inside the conductor which are parallel to the surface. What is the characteristic penetration depth of the field?

b) What is the ratio of the magnetic field amplitude to the electric field amplitude inside the metal?

c) What is the power per unit area transmitted into the metal?

3.47   Wave Attenuation (Stony Brook)

Consider a medium with nonzero conductivity σ ($\mathbf{J} = \sigma \mathbf{E}$ gives the current density) and no net charge ($\rho = 0$).

a) Write the set of Maxwell's equations appropriate for this medium.

b) Derive the wave equation for $\mathbf{E}$ (or $\mathbf{B}$) in this medium,

$$\left[\nabla \times (\nabla \times \mathbf{V}) = \nabla (\nabla \cdot \mathbf{V}) - \nabla^2 \mathbf{V} \right]$$

c) Consider a monochromatic wave moving in the $+x$ direction with E_y (or E_z or B_y or B_z) given by

$$E_y = \psi = \psi_0 e^{i(kx - \omega t)}$$

Show that this wave has an amplitude which decreases exponentially; find the attenuation length (skin depth).

d) For sea water ($\sigma \approx 5$ mho/m, or $4.5 \cdot 10^{10} \mathrm{s}^{-1}$ in cgs units), and using radio waves of long wavelength $\omega = 5 \cdot 10^5 \mathrm{s}^{-1}$, calculate the attenuation length. (Why is it hard to communicate with submerged submarines?) You can take for sea water $\varepsilon = 1, \mu = 1$.

3.48   Electrons and Circularly Polarized Waves (Boston)

a) An electron (mass m, charge $-e$) is subject to the elastic force $\mathbf{F} = -m\omega_0^2 \mathbf{r}$. A static, uniform magnetic field $\mathbf{B}$ in the positive z direction is also present. The electron is set in forced motion by one of the circularly polarized electromagnetic waves

$$\mathbf{E} = E_0 \left[\cos(kz - \omega t)\hat{\mathbf{i}} + \cos(kz - \omega t + (\varepsilon \pi/2))\hat{\mathbf{j}} \right]$$

($\varepsilon = -1$ positive helicity, $\varepsilon = 1$ negative helicity). Find the coordinates of the electron as functions of time.

Hints: Approximate $\mathbf{E}$ by its value for $z = 0$. The use of the variables $\xi = x + iy$ and $\eta = x - iy$ will simplify your calculations.

b) Consider circularly polarized electromagnetic waves propagating in the direction of a static magnetic field $\mathbf{B}$ in a medium consisting of elastically bound electrons (N per unit volume). The elementary theory of dispersion leads to the formula

$$n_+^2 - n_-^2 = \frac{4\pi e^2 N}{m} \left[\frac{1}{\omega_0^2 - \omega^2 + eB\omega/mc} - \frac{1}{\omega_0^2 - \omega^2 - eB\omega/mc} \right]$$

(P.3.48.1)

for the refractive indices (n_+, n_-) of circularly polarized waves of (positive, negative) helicity.
Without proving this formula, infer from it that the plane of polarization of a linearly polarized wave is rotated through the angle

$$\chi \approx \frac{4\pi e^2 N}{m} \frac{1}{(\omega_0^2 - \omega^2)^2} \frac{eB\omega^2}{mc^2} \frac{l}{2n}$$

after propagating through a length l in the medium when a magnetic field in the direction of propagation is present.

Hint: Represent a linearly polarized wave as a sum of circularly polarized waves of opposite helicities.

3.49   Classical Atomic Spectral Line (Princeton, Wisconsin-Madison)

Consider the classical theory of the width of an atomic spectral line. The "atom" consists of an electron of mass m and charge e in a harmonic oscillator potential. There is also a frictional damping force, so the equation of motion for the electron is

$$m\ddot{\mathbf{x}} + m\omega_0^2 \mathbf{x} + \gamma \dot{\mathbf{x}} = 0$$

(P.3.49.1)

a) Suppose at time $t = 0$, $\mathbf{x} = \mathbf{x}_0$, and $\dot{\mathbf{x}} = 0$. What is the subsequent motion of the electron? A classical electron executing this motion would emit electromagnetic radiation. Determine the intensity $I(\omega)$

of this radiation as a function of frequency. (You need not calculate the absolute normalization of $I(\omega)$, only the form of the ω dependence of $I(\omega)$. In other words, it is enough to calculate $I(\omega)$ up to a constant of proportionality.) Assume $\gamma/m \ll \omega_0$.

b) Now suppose the damping force $\gamma \dot{x}$ is absent from (P.3.49.1) and that the oscillation is damped only by the loss of energy to radiation (an effect which has been ignored above). The energy U of the oscillator decays as $U_0 e^{-\Gamma t}$. What, under the above assumptions, is Γ? (You may assume that in any one oscillation the electron loses only a very small fraction of its energy.)

c) For an atomic spectral line of 5000 Å, what is the width of the spectral line, in angstroms, as determined from the calculation of (b)? About how many oscillations does the electron make while losing half its energy? Rough estimates are enough.

3.50 Lifetime of Classical Atom (MIT, Princeton, Stony Brook)

At a time $t = 0$, the electron orbits a classical hydrogen atom at a radius a_0 equal to the first Bohr radius. Derive an expression for the time it takes for the radius to decrease to zero due to radiation. Assume that the energy loss per revolution is small compared to the total energy of the atom.

3.51 Lorentz Transformation of Fields (Stony Brook)

a) Write down the Lorentz transformation equations relating the space–time coordinates of frames K and K', where K' moves with velocity v relative to K. (Take $\mathbf{v}$ to point along a coordinate axis for simplicity.) Explicitly define your 4-vector conventions.

b) Use the fact that the electromagnetic field components $\mathbf{E}$ and $\mathbf{B}$ form an antisymmetric tensor to show that

$$\mathbf{E}'_\parallel = \mathbf{E}_\parallel \qquad \mathbf{E}'_\perp = \gamma \left(\mathbf{E}_\perp + \frac{\mathbf{v}}{c} \times \mathbf{B} \right)$$

$$\mathbf{B}'_\parallel = \mathbf{B}_\parallel \qquad \mathbf{B}'_\perp = \gamma \left(\mathbf{B}_\perp - \frac{\mathbf{v}}{c} \times \mathbf{E} \right)$$

where $\gamma = 1/\sqrt{1 - (v/c)^2}$ and the subscripts label directions parallel and perpendicular to $\mathbf{v}$.

c) Consider the particular case of a point charge q and recover an appropriate form of the law of Biot and Savart for small v.

3.52   Field of a Moving Charge (Stony Brook)

A charged particle with charge q_1 moves with constant velocity $\mathbf{v}$ along the
z-axis (see Figure P.3.52). Its potentials are

$$\phi = \frac{q_1}{s} \qquad \mathbf{A} = \frac{\mathbf{v}}{c}\phi$$

where $s = \left[\left(1 - \beta^2\right)\left(x^2 + y^2\right) + \left(z - vt\right)^2\right]^{1/2}$ and $\beta = v/c$

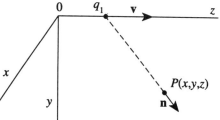

Figure P.3.52

a) Show that ϕ and $\mathbf{A}$ satisfy the Lorentz condition

$$\frac{1}{c}\frac{\partial \phi}{\partial t} + \nabla \cdot \mathbf{A} = 0$$

b) Calculate the fields $\mathbf{E}$ and $\mathbf{B}$ at point $P(x, y, z)$ and time t. Show
first that $\mathbf{B} = (\mathbf{v}/c) \times \mathbf{E}$, and calculate $\mathbf{E}$ explicitly. Show that $\mathbf{E}$ is
parallel to $\mathbf{n}$.

c) Assume at P a second particle with charge q_2 moving with the same
velocity $\mathbf{v}$ as the first. Calculate the force on q_2.

3.53   Retarded Potential of Moving Line Charge (MIT)

An infinitely long insulating filament with linear charge density λ lies at
rest along the z-axis (see Figure P.3.53).

a) Find the electrostatic field E_r at a point P a distance x_0 away from
the origin along the x-axis.

b) At $t = 0$, the wire suddenly starts to move with constant velocity v
in the positive z direction. Assuming the wire is infinitely thin, write
down an expression for the current density $\mathbf{J}$ arising from the motion.
Using the formula for the retarded potential

$$\mathbf{A}(\mathbf{x}, t) = \frac{1}{c}\int d^3x' \frac{\mathbf{J}\left(\mathbf{x}', t - |\mathbf{x} - \mathbf{x}'|/c\right)}{|\mathbf{x} - \mathbf{x}'|}$$

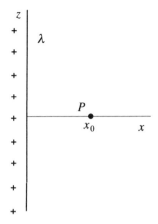

Figure P.3.53

calculate $A_z(x_0, t)$. Give its value for $t > x_0/c$ and for $t < x_0/c$.

c) Because of cylindrical symmetry, you really know $A_z(\rho, t)$ with ρ the radial coordinate in cylindrical coordinates. Find $\mathbf{B}(\rho, t)$ as $t \to \infty$. Does your value agree with your intuitive expectation from Ampère's law?

Hint: A useful integral is

$$\int \frac{dx}{\sqrt{x^2 + a^2}} = \ln \left| \frac{x + \sqrt{x^2 + a^2}}{a} \right|$$

3.54 Orbiting Charges and Multipole Radiation (Princeton, Michigan State, Maryland)

Charges $+q$ and $-q$ a distance d apart orbit around each other in the x–y plane $(z = 0)$ at frequency ω $(d \ll c/\omega)$ (see Figure P.3.54).

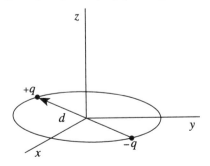

Figure P.3.54

a) The emitted radiation is primarily confined to one multipole. Which one?
b) What is the angular distribution of the radiated power?
c) What is the total power radiated?
d) The plane $z = -b$ is now filled with a perfect conductor ($b \ll c/\omega$). What multipole radiates now?

3.55 Electron and Radiation Reaction (Boston)

The equation of motion for a particle of mass m and charge q in electric and magnetic fields $\mathbf{E}$ and $\mathbf{B}$, including the radiation reaction force, is

$$m\dot{\mathbf{v}} = q\left(\mathbf{E} + \frac{1}{c}\mathbf{v} \times \mathbf{B}\right) + \frac{2q^2}{3c^3}\ddot{\mathbf{v}}$$

a) Assuming that the radiative reaction term is very small compared to the Lorentz force and that $v \ll c$, find an approximate expression for the radiative reaction force in terms of $\mathbf{E}$ and $\mathbf{B}$.
b) A plane electromagnetic wave propagates in the z direction. A free electron is initially at rest in this wave. Under the assumptions of (a), calculate the time-averaged radiative reaction force on the electron (magnitude and direction). What result would you obtain for a positron?
c) Rederive the reaction force by considering the momentum acquired by the electron in the process of forced emission of radiation. Use the Thomson cross section σ

$$\sigma = \frac{\text{scattered power}}{\text{incident power/unit area}} = \frac{8\pi}{3}\left(\frac{e^2}{mc^2}\right)$$

3.56 Radiation of Accelerating Positron (Princeton, Colorado)

A nonrelativistic positron of charge e and velocity v_1 ($v_1 \ll c$) impinges head-on on a fixed nucleus of charge Ze (see Figure P.3.56). The positron, which is coming from far away (∞), is decelerated until it comes to rest and then is accelerated again in the opposite direction until it reaches a terminal velocity v_2. Taking radiation loss into account (but assuming it is small), find v_2 as a function of v_1 and Z.

Figure **P.3.56**

3.57 Half-Wave Antenna (Boston)

Consider the half-wave antenna shown in Figure P.3.57. The current distribution shown as a broken line is $I = I_0 \cos(2\pi z/\lambda) \cos(\omega t)$.

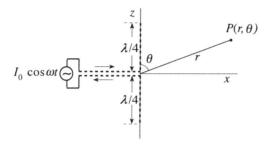

Figure **P.3.57**

a) Find the vector potential in the radiation zone due to the complex current $I = I_0 \cos(2\pi z/\lambda) \exp(i\omega t)$.

b) Find the electric field **E** and the magnetic induction **B** in the radiation zone.

c) Show that the time-averaged power radiated per unit solid angle is

$$\frac{dP}{d\Omega} = \frac{I_0^2}{2\pi c}\frac{\cos^2((\pi/2)\cos\theta)}{\sin^2\theta}$$

Hint:

$$\frac{dP}{d\Omega} = \frac{c}{8\pi}\mathrm{Re}\left[r^2\mathbf{n}\cdot(\mathbf{E}\times\mathbf{B}^*)\right]$$

3.58 Čerenkov Radiation (Stony Brook)

Čerenkov radiation is an electromagnetic shock wave caused by a charged particle moving with a velocity v which is faster than the velocity of light c/n in a medium with index of refraction n.

a) Show that the shock wave is emitted at an angle θ_c relative to the particle direction, where

$$\cos\theta_c = \frac{1}{n\beta} \qquad (\beta \equiv \frac{v}{c})$$

b) Show that a spherical mirror with radius of curvature R will focus this shock wave onto a ring in the focal plane of the mirror.

c) Find the radius of the ring.

3.59   Stability of Plasma (Boston)

Consider an idealized ion beam of radius R and length much longer than R.

  a) Show that an individual ion at the periphery of this beam is subject to the net outward force

$$F = \frac{2IQ}{Rv}\left(1 - \frac{v^2}{c^2}\right)$$

  where I is the beam current, Q is the charge of each ion, and v is the velocity of the ions. Assume that the charge and current densities have cylindrical symmetry.

  b) The beam diverges because the electrostatic force prevails on the magnetic force that tends to concentrate the beam along its axis ("pinch" effect). Show that the rate of increase of the beam radius

$$\frac{dr}{dt} = \sqrt{\frac{4IQ}{Mv}\left(1 - \frac{v^2}{c^2}\right)\ln(r/R)}$$

3.60   Charged Particle in Uniform Magnetic Field (Princeton)

A nonrelativistic charged particle is orbiting in a uniform magnetic field of strength H_0 at the center of a large solenoid. The radius of the orbit is R_0. The field is changed slowly to H_1. What is the new radius R_1 of the orbit? If the field is suddenly changed back to H_0, what is the final radius R_2?

3.61   Lowest Mode of Rectangular Wave Guide (Princeton, MIT, Michigan State)

Consider a rectangular wave guide, infinitely long in the z direction, with a width (x direction) of a and a height (y direction) of b ($a > b$) (see Figure P.3.61). The walls are perfect conductors.

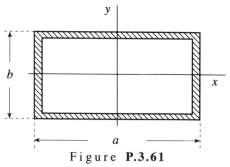

Figure **P.3.61**

a) What are the boundary conditions on the components of **B** and **E** at the walls?

b) Write the wave equation which describes the **E** and **B** fields of the lowest mode.

Hint: The lowest mode has the electric field in the y direction only.

c) For the lowest mode that can propagate, find the phase velocity and the group velocity.

d) The possible modes of propagation separate naturally into two classes. What are these two classes and how do they differ physically?

3.62 TM Modes in Rectangular Wave Guide (Princeton)

A rectangular wave guide of sides $a = 7.21$ cm and $b = 3.40$ cm is used in the transverse magnetic (TM) mode (see Figure P.3.61). TM modes are modes in which the magnetic field is perpendicular to the direction of propagation, here z. Assume that the walls are perfect conductors.

a) By calculating the lowest cutoff frequency, determine whether TM radiation of angular frequency $\omega = 6.1 \times 10^{10} s^{-1}$ will propagate in the wave guide.

b) What is the dispersion relation (i.e., the relationship between ω and the wavevector k) for this guide.

c) Find the attenuation length, i.e., the distance over which the power drops to e^{-1} of its starting value, for a frequency ω that is *half* the cutoff frequency.

3.63 Betatron (Princeton, Moscow Phys-Tech, Colorado, Stony Brook (a))

Consider the motion of electrons in an axially symmetric magnetic field. Suppose that at $z = 0$ (the "median plane") the radial component of the

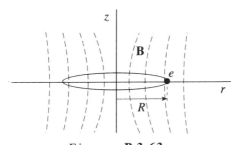

Figure **P.3.63**

magnetic field is 0, so $\mathbf{B}(z=0) = B(r)\hat{z}$. Electrons at $z = 0$ then follow a circular path of radius R (see Figure P.3.63).

a) What is the relationship between the electron momentum p and the orbit radius R?

In a betatron, electrons are accelerated by a magnetic field which changes with time. Let $\langle B \rangle$ equal the average value of the magnetic field over the plane of the orbit (within the orbit), i.e.,

$$\langle B \rangle = \frac{\Phi}{\pi R^2}$$

where Φ is the magnetic flux through the orbit. Let B_0 equal $\mathbf{B}(r{=}R, z{=}0)$.

b) Suppose $\langle B \rangle$ is changed by an amount $\langle \Delta B \rangle$ and B_0 is changed by ΔB_0. How must $\langle \Delta B \rangle$ be related to ΔB_0 if the electrons are to remain at radius R as their momentum is increased?

c) Suppose the z component of the magnetic field near $r = R$ and $z = 0$ varies with r as $B_z(r) = B_0(R)(R/r)^n$. Find the equations of motion for small departures from the equilibrium orbit in the median plane. There are two equations, one for small vertical changes and one for small radial changes. Neglect any coupling between radial and vertical motion.

d) For what range of n is the orbit stable against both vertical and radial perturbations?

3.64   Superconducting Frame in Magnetic Field (Moscow Phys-Tech)

A superconducting square rigid frame of mass m, inductance L, and side a is cooled down (in a magnetic field) to a temperature below the critical temperature. The frame is kept horizontal (parallel to the x–y plane) and constrained to move in the z direction in a nonuniform but constant

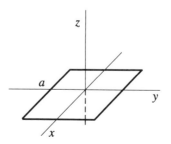

Figure **P.3.64**

magnetic field described by a vector potential $\mathbf{A} = (-B_0 y, \alpha xz, 0)$ and a uniform gravitational field given by the acceleration g. The thickness of the frame is much smaller than a (see Figure P.3.64). Initially, the frame is at rest, with its center coinciding with the origin. Find the equations of motion of the frame and solve for the position of the frame as a function of time.

3.65 Superconducting Sphere in Magnetic Field (Michigan State, Moscow Phys-Tech)

A superconducting (Type I) spherical shell of radius R is placed in a uniform magnetic field $\mathbf{B}_0$ ($\mathbf{B}_0 \ll \mathbf{H}_c$, $\mathbf{H}_c$, the critical field). Find

a) the magnetic field everywhere outside the shell
b) the surface current density

Hint: Inside, $\mathbf{B} = 0$.

3.66 London Penetration Depth (Moscow Phys-Tech)

The model used by the brothers F. and H. London suggests that free electrons in a superconductor can be divided into two types: normal, with a density n_n, and superconducting, with a density n_s (where $n_s + n_n = n$, the density of free electrons). On the surface of the superconductor flows a current with density $\mathbf{J}_s$.

a) Using classical arguments, obtain the equation for the electric field $\mathbf{E}$, where

$$\mathbf{E} = \frac{d}{dt}(\Lambda \mathbf{J}_s) \qquad (\text{P.3.66.1})$$

where $\Lambda = m/(n_s e^2)$. Here, m and e are the mass and charge of the electron. Again using classical arguments, write the kinetic energy of superconducting electrons in the form

$$\mathcal{E}_{\text{kin}} = \frac{n_s m v_s^2}{2}$$

Adding the magnetic field energy, find a minimum of the free energy $\mathcal{F}_s(\mathbf{h})$ to obtain a second equation

$$\mathbf{h} + \lambda_L^2 \nabla \times \nabla \times \mathbf{h} = 0 \qquad (\text{P.3.66.2})$$

where $\mathbf{h}$ is a microscopic magnetic field inside the superconductor and λ_L is the London penetration depth.

b) Solve (P.3.66.2) for the boundary between vacuum and a space half-filled with superconductor, with an external field H_0 parallel to the boundary, and estimate $\lambda_L(0)$ for a typical metal superconductor at zero temperature, assuming $n_s(0) = n$.

3.67   Thin Superconducting Plate in Magnetic Field (Stony Brook)

A very long, thin plate of thickness d is placed in a uniform magnetic field $\mathbf{H}_0$ parallel to the surface of the plate (see Figure P.3.67).

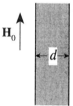

Figure **P.3.67**

a) Find the distribution of the magnetic field and the current inside the plate.

b) Consider two limiting cases $d \gg \lambda_L, d \ll \lambda_L$, and draw a picture of the distributions in these two cases (λ_L − London penetration depth).

SOLUTIONS

Mechanics

1.1 Falling Chain (MIT, Stanford)

The reading of the scale consists of two parts: the weight of the chain accumulated and the impulse per unit time imparted by the chain colliding with the scale. The first part is clearly $W_1 = Mgx/L$. The velocity of the links at the instant of hitting the scale is found from $v^2 = 2gx$. The second component of the force exerted on the scale equals

$$\frac{dp}{dt} = \frac{d(mv)}{dt} = v\frac{dm}{dt} = v\frac{M}{L}v = 2Mg\frac{x}{L}$$

The total force on the scale during the fall of the chain is therefore

$$Mg\frac{x}{L} + 2Mg\frac{x}{L} = 3Mg\frac{x}{L}$$

1.2 Cat and Mouse Tug of War (Moscow Phys-Tech, MIT)

a) The straightforward solution is given in (b). However, to answer the first part, we can use dimensional analysis. The force f applied by the mice may depend only on the dimensionless coefficient of friction μ, the cylinder diameter d, and the force F :

$$f \propto G(\mu)\, d^\alpha F^\beta$$

where $G(\mu)$ is some function of μ. It is obvious that it is impossible to satisfy this equation unless $\alpha = 0$, so the force f does not depend on the diameter of the cylinder.

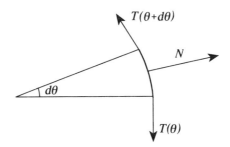

Figure   S.1.2

b)   Consider an element of rope between angles θ and $\theta + d\theta$. The difference in tension dT between its ends equals $dT = \mu N = \mu T d\theta$, where N is the normal force (see Figure S.1.2). This equation has a solution: $T = \text{const} \cdot e^{\mu\theta}$. At $\theta = 0$, $T = f$, and at $\theta_0 = \pi/3$, $T = F$, so

$$F = f e^{\mu\theta_0} \tag{S.1.2.1}$$

From equation (S.1.2.1) we obtain the same result as determined in part (a), i.e., the force does not depend on the diameter of the cylinder. Similarly, if we have just one mouse, the minimum angle θ_1 of wrapping is found from

$$F = f' e^{\mu\theta_1} \tag{S.1.2.2}$$

where $f' = f/10 = F/100$ is the force necessary for one mouse to keep the rope from slipping. From equation (S.1.2.1), we can also find the coefficient of friction μ :

$$\mu = \frac{1}{\theta_0} \ln\left(\frac{F}{f}\right)$$

Then, from equation (S.1.2.2),

$$\theta_1 = \frac{1}{\mu} \ln\left(\frac{F}{f'}\right) = \theta_0 \frac{\ln\left(F/f'\right)}{\ln\left(F/f\right)} = \theta_0 \log_{F/f}\left(\frac{F}{f'}\right) = \theta_0 \log_{10} 100 = 2\theta_0 = \frac{2\pi}{3}$$

1.3   Cube Bouncing off Wall (Moscow Phys-Tech)

There are two forces acting on the cube. One is the normal reaction $N(t)$, perpendicular to the wall, and the other is the force of friction $F_{fr}(t)$, parallel to the wall (see Figure S.1.3). We expect that, as a result of the collision, the cube's velocity $\mathbf{v}$ will change to $\mathbf{v}'$. In the direction perpendicular to

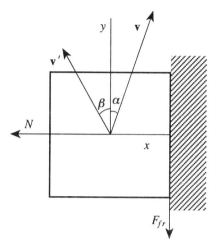

Figure   S.1.3

the wall, the collision is elastic, i.e., the velocity in the $\hat{x}$ direction merely
changes sign: $v'_x = -v_x = -v \sin \alpha$. Therefore, the momentum changes by
$-2mv \sin \alpha$ in the $\hat{x}$ direction. This change is due to the normal reaction
$N(t)$. So, according to Newton's second law:

$$\Delta p_x = - \int\limits_0^\tau N(t) \, dt = -2mv \sin \alpha \qquad (S.1.3.3)$$

where τ is the collision time. If there were no friction, the parallel velocity component would not change and the angle would remain the same.
However, in the actual case, the $\hat{y}$ component changes and

$$\Delta p_y = - \int\limits_0^{\min\{\tau',\tau\}} F_{fr}(t) \, dt = - \int\limits_0^{\min\{\tau',\tau\}} \mu N(t) \, dt \qquad (S.1.3.4)$$

Here τ' is the time at which the velocity v_y goes to zero. So, from (S.1.3.4)

$$mv'_y = mv_y - \int\limits_0^{\min\{\tau',\tau\}} \mu N(t) \, dt \qquad (S.1.3.5)$$

First assume that $\tau < \tau'$, i.e., v'_y is not zero. Then we have from (S.1.3.3)

$$2mv \sin \alpha = \int\limits_0^\tau N(t) \, dt \qquad (S.1.3.1)$$

and

$$mv'_y = mv\cos\alpha - \mu \int_0^\tau N(t)\,dt = mv\cos\alpha - 2\mu mv\sin\alpha \qquad \text{(S.1.3.2)}$$

or

$$v'_y = v(\cos\alpha - 2\mu\sin\alpha) \qquad \text{(S.1.3.3)}$$

Therefore, the angle β between the velocity $\mathbf{v}'$ and the wall is given by

$$\tan\beta = \frac{|v'_x|}{v'_y} = \frac{v\sin\alpha}{v(\cos\alpha - 2\mu\sin\alpha)} = \frac{\tan\alpha}{1 - 2\mu\tan\alpha} \qquad \text{(S.1.3.4)}$$

If $1 - 2\mu\tan\alpha = 0$, then $\tan\alpha = (1/2\mu)$ and $\tan\beta$ goes to infinity, which corresponds to $\beta = \pi/2$. Now if $1 - 2\mu\tan\alpha < 0$ (v_y goes to zero before the collision ends; v'_y cannot become negative), then $\tau > \tau'$, and the cube will leave perpendicular to the wall. Therefore,

$$\beta = \begin{cases} \tan^{-1}\left(\dfrac{\tan\alpha}{1 - 2\mu\tan\alpha}\right), & \tan\alpha < \dfrac{1}{2\mu} \\[3ex] \dfrac{\pi}{2}, & \tan\alpha \geq \dfrac{1}{2\mu} \end{cases}$$

1.4 Cue-Struck Billiard Ball (Rutgers, Moscow Phys-Tech, Wisconsin-Madison (a))

a) Introduce a frame of reference with the origin at the center of the ball (see Figure S.1.4a). Since the direction of the force is toward the center of the ball there is no torque at $t = 0$. (We consider a very short pulse.) So one can define the initial conditions of the ball's movement from the

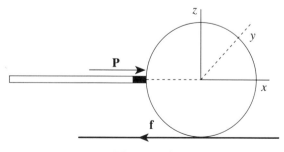

Figure S.1.4a

---

equations

$$m\mathbf{v}(0) = \mathbf{P}$$

$$I\boldsymbol{\omega}(0) = 0 \tag{S.1.4.1}$$

where $\mathbf{v}(0)$ and $\boldsymbol{\omega}(0)$ are the velocity and angular velocity at $t=0$ and I is the moment of inertia of the ball $I = (2/5)\,mR^2$. So we have $\mathbf{v}(0) = \mathbf{P}/m$ and $\boldsymbol{\omega}(0) = 0$. Subsequent motion of the ball will be given by

$$\mathbf{v}(t) = \mathbf{v}(0) + \frac{\mathbf{f}}{m}t$$

$$\boldsymbol{\omega}(t) = \boldsymbol{\omega}(0) + \frac{\mathbf{R}\times\mathbf{f}}{I}t \tag{S.1.4.2}$$

where $\mathbf{f}$ is a friction force $\mathbf{f} = -\mu mg\hat{\mathbf{x}}$ and $\boldsymbol{\omega} = \omega\hat{\mathbf{y}}$. In our geometry, (S.1.4.2) may be rewritten in the form

$$v(t) = v(0) - \frac{f}{m}t$$

$$\omega(t) = \omega(0) + \frac{fR}{I}t \tag{S.1.4.3}$$

The ball will roll without slipping when $\mathbf{v} = \boldsymbol{\omega}\times\mathbf{R}$ or $v = \omega R$. Using (S.1.4.3), we obtain the time t_R when pure rolling motion begins:

$$t_R = \frac{v(0) - \omega(0)R}{(f/m)(1+mR^2/I)} = \frac{v(0)}{(f/m)(1+mR^2/I)} = \frac{2}{7}\frac{P}{f}$$

The final speed v_f of the center of mass of the ball is given by

$$v_f = v(0) - \frac{f}{m}t_R = \frac{P}{m}\left(1-\frac{2}{7}\right) = \frac{5}{7}\frac{P}{m}$$

b) Using (S.1.4.1) from part (a) for the initial conditions, we obtain

$$mv(0) = P$$

$$I\omega(0) = hP$$

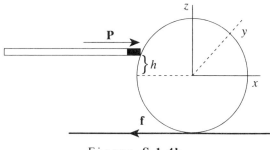

Figure S.1.4b

Again using the condition $v(0) = \omega(0)R$ for rolling without slipping, we have (see Figure S. 1.4b)

$$\frac{P}{m} = \frac{hP}{I}R$$

and

$$h = \frac{2}{5}R$$

1.5 Stability on Rotating Rollers (Princeton)

Hint: Consider a rod, e.g., a pencil, supported by one finger of each hand. First put your fingers as far apart as possible and then move them until they touch. Where do they meet? Now put your fingers together and place the rod with its center of mass at this point and move your fingers apart. What happens now?

a) Let us orient the x-coordinate positive to the right (see Figure S.1.5a). Then we can write equations for forces and torques relative to the center

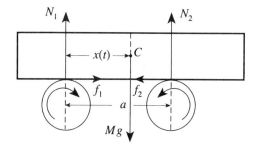

Figure S.1.5a

of mass of the rod:

$$N_1 + N_2 = Mg \tag{S.1.5.1}$$

$$N_1 x = N_2 (a - x) \tag{S.1.5.2}$$

$$M\ddot{x} = f_1 - f_2 \tag{S.1.5.3}$$

where N_1 and N_2 are normal forces and f_1 and f_2 are friction forces at the first and second rollers, respectively.

$$f_1 = \mu N_1 \qquad f_2 = \mu N_2 \tag{S.1.5.4}$$

From (S.1.5.1) and (S.1.5.2) we get

$$N_1 = Mg \left(1 - \frac{x}{a}\right)$$
$$\tag{S.1.5.5}$$
$$N_2 = Mg\frac{x}{a}$$

Substituting (S.1.5.5) and (S.1.5.4) into (S.1.5.3) results in the differential equation

$$\ddot{x} + \frac{2\mu g}{a}x - \mu g = 0$$

Letting $\omega^2 = 2\mu g/a$ gives $\ddot{x} + \omega^2 x = \mu g$. The solution of this equation is

$$x = A\cos(\omega t + \alpha) + \frac{a}{2}$$

where α is an arbitrary phase. Taking into account the initial conditions $x(0) = x_0$ and $\dot{x}(0) = 0$ leads to the solution

$$x(t) = \left(x_0 - \frac{a}{2}\right)\cos \omega t + \frac{a}{2}$$

corresponding to simple harmonic motion.

b) Now consider another case (see Figure S.1.5b). The equations are quite similar:

$$N_1 + N_2 = Mg \tag{S.1.5.1'}$$

$$N_1 x = N_2 (a - x) \tag{S.1.5.2'}$$

$$M\ddot{x} = f_2 - f_1 = \mu (N_2 - N_1) \tag{S.1.5.3'}$$

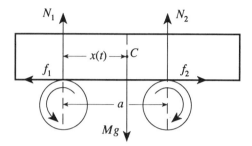

Figure S.1.5b

Again from (S.1.5.1′) and (S.1.5.2′) we can get $\ddot{x} - \omega^2 x = -\mu g$, $\omega^2 = 2\mu g/a$. The solution of this equation is

$$x\left(t\right) = \left(x_0 - \frac{a}{2}\right)\cosh\omega t + \frac{a}{2}.$$

This means that the motion is not bound within the length of the rod. Even if you place the rod in the middle of the rollers, the equilibrium will not be stable.

1.6   Swan and Crawfish (Moscow Phys-Tech)

First consider the method most likely suggested by Swan. Ignorant of the law of gravitational attraction, he does not have to apply a strictly horizontal force, and although the horizontal projection of the force is smaller, the friction is also smaller. Let us assume that he applies the force at the center of mass of the dresser and at an angle α to the horizontal (see Figure S.1.6a). Then we can write, for the normal force $\mathbf{N}$ exerted by the floor on

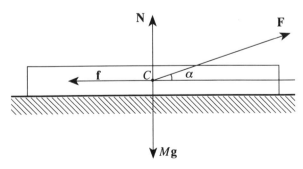

Figure S.1.6a

the dresser,

$$N = Mg - F\sin\alpha$$

To move the dresser, Swan needs to apply a force **F** such that its horizontal projection is larger than the friction force $\mathbf{f} = \mu\mathbf{N}$:

$$F\cos\alpha > \mu\,(Mg - F\sin\alpha)$$

or

$$F > \frac{\mu Mg}{\cos\alpha + \mu\sin\alpha} \qquad\qquad\text{(S.1.6.1)}$$

It is easy to check that $\tan\alpha = \mu$ corresponds to the maximum of the denominator in (S.1.6.1) and therefore to a minimum force F. Using $\cos\alpha = 1/\sqrt{1 + \tan^2\alpha}$, we have

$$F_{\min} = \frac{\mu Mg}{\sqrt{1 + \mu^2}} \approx 660\text{ N}$$

Therefore, the force F should be

$$F > 660\text{ N}$$

So Swan, who can apply a force of 700 N, will be able to move the dresser alone.

Crawfish, being somewhat more earthbound, is likely to suggest another method of moving the piece of furniture. He will apply a horizontal force, but not to the center of mass of the dresser; rather, to one of its ends. The dresser will start to rotate; however, the center of rotation R will not coincide with the center of mass of the dresser (see Figure S.1.6b). So, after one rotation of $180°$, Crawfish will have moved the dresser by $l - 2x$, where

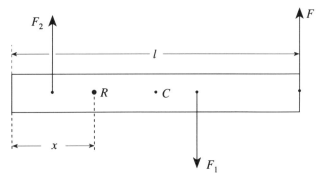

Figure **S.1.6b**

l is the total length of the dresser. The frictional forces on each of the two parts are F_1 and F_2. They are proportional to the weight of the two parts:

$$F_1 = \frac{l - x}{l} \mu M g \qquad \text{(S.1.6.2)}$$

$$F_2 = \frac{x}{l} \mu M g \qquad \text{(S.1.6.3)}$$

Since the torque relative to the point R is zero, we have

$$F(l - x) - F_1 \left(\frac{l - x}{2} \right) - F_2 \frac{x}{2} = 0$$

Taking F_1 and F_2 from (S.1.6.2) and (S.1.6.3), we obtain the relation between F and x :

$$F = \frac{2x^2 - 2lx + l^2}{2l(l - x)} \mu M g$$

The minimum force F occurs when

$$x = \left(1 - \frac{\sqrt{2}}{2} \right) l \qquad F_{\min} = \left(\sqrt{2} - 1 \right) \mu M g \approx 305 \text{ N}$$

So, Crawfish is also able to move the dresser by his method.

1.7 Mud from Tire (Stony Brook)

Mud flying from different points on the tire will rise to different heights, depending on the initial height and angle of ejection. Introducing an angle α and the height h of the point of ejection O above the equator of the tire (see Figure S.1.7), we can write using energy conservation

$$mgh_0 = \frac{mv_y^2}{2} = \frac{mv^2 \cos^2 \alpha}{2}$$

$$h_0 = \frac{v^2 \cos^2 \alpha}{2g}$$

where h_0 is the height to which the mud rises above O, and $v = \omega R$ is the speed of the rim of the wheel. The height H above the ground will be

$$H = R + h + h_0 = R + R\sin \alpha + \frac{v^2 \cos^2 \alpha}{2g}$$

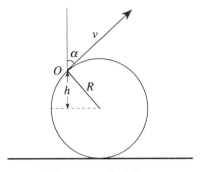

Figure   S.1.7

Now find the maximum height by setting the derivative of H with respect to α equal to 0:

$$\frac{dH}{d\alpha} = 0 = R\cos\alpha - \frac{v^2}{g}\cos\alpha\sin\alpha$$

There are two solutions of this equation. First,

$$\cos\alpha = 0 \qquad\qquad \alpha = \frac{\pi}{2} \qquad\qquad (\text{S.1.7.1})$$

This case yields a maximum only when $v^2 \leq gR$, and the highest point of the wheel is the maximum height. But here $v^2 > gR$, so we will consider the other case:

$$\sin\alpha_0 = \frac{gR}{v^2} \qquad\qquad (\text{S.1.7.2})$$

The height becomes

$$H = R + R\sin\alpha_0 + \frac{v^2}{2g}\left(1 - \sin^2\alpha_0\right) = R + \frac{v^2}{2g} + \frac{gR^2}{2v^2}$$

We can check that

$$\left.\frac{d^2H}{d\alpha^2}\right|_{\alpha_0} = -R\sin\alpha_0 - \frac{v^2}{g}\cos^2\alpha_0 + \frac{v^2}{g}\sin^2\alpha_0$$

$$= -\frac{v^2}{g} + \frac{gR^2}{v^2} < 0 \ (\text{maximum})$$

1.8 Car down Ramp up Loop (Stony Brook)

a) Since there is no friction, we have from energy conservation

$$mgh_0 = mgh + \frac{1}{2}mv^2$$

where v is the velocity of the car and m is its mass (see Figure S.1.8). At
any point of the ramp, the normal force of the ramp on the car should

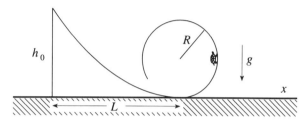

Figure   S.1.8

be nonzero; otherwise, the car will leave the ramp. Obviously, the critical
point is the top of the loop, $x = L$, where the velocity is minimal and
gravity and centrifugal forces are antiparallel. For this point,

$$mgh_0 = 2mgR + \frac{1}{2}mv^2 \qquad (S.1.8.1)$$

The minimum height h_0 corresponds to a velocity $v = v_{\min}$ at this point,
enough to provide a centrifugal acceleration equal to the gravitational acceleration:

$$\frac{v_{\min}^2}{R} = g$$

Substituting this into (S.1.8.1) yields

$$mgh_0 = 2mgR + \frac{1}{2}mgR = \frac{5}{2}mgR$$

$$h_0 = \frac{5}{2}R$$

b) Consider a point on the ramp $0 < x < L$. The velocity at this point is
defined by

$$\frac{mv^2}{2} = mgh_0 - mgh(x) \qquad (S.1.8.2)$$

where v^2 may be written

$$v^2 = \left(\dot{x}^2 + \dot{y}^2\right) = \left[\dot{x}^2 + (dy/dx)^2\, \dot{x}^2\right] = \left[1 + (dh/dx)^2\right]\dot{x}^2 \qquad (S.1.8.3)$$

where the slope of the curve dy/dx is given by the derivative of the height function dh/dx. Substituting (S.1.8.3) into (S.1.8.2) results in

$$\dot{x} = \sqrt{\frac{2g\left[h_0 - h\left(x\right)\right]}{1 + (dh/dx)^2}}$$

c) Now consider $h\left(x\right) = h_0\left[1 - \sin\left(\pi x/2L\right)\right]$. Rewrite the solution to (b) as

$$dt = dx\sqrt{\frac{1 + (dh/dx)^2}{2g\left[h_0 - h\left(x\right)\right]}}$$

The time T to travel to $x = L$ can be expressed by the integral

$$T = \int_0^L dx\sqrt{\frac{1 + (dh/dx)^2}{2g\left[h_0 - h(x)\right]}} \qquad \frac{dh}{dx} = -\frac{\pi h_0}{2L}\cos\left(\frac{\pi x}{2L}\right)$$

$$T = \int_0^L dx\sqrt{\frac{1 + (\pi h_0/2L)^2\cos^2\left(\pi x/2L\right)}{2gh_0\sin\left(\pi x/2L\right)}}$$

Letting $\xi \equiv \pi x/2L$ and $a \equiv \pi h_0/2L$, we obtain

$$T = \frac{L}{\sqrt{gh_0}}\,\frac{\sqrt{2}}{\pi}\int_0^{\pi/2} d\xi\sqrt{\frac{1 + a^2\cos^2\xi}{\sin\xi}} = \frac{L}{\sqrt{gh_0}}f\left(a\right)$$

where

$$f\left(a\right) \equiv \frac{\sqrt{2}}{\pi}\int_0^{\pi/2} d\xi\sqrt{\frac{1 + a^2\cos^2\xi}{\sin\xi}}$$

In the limiting case of $h_0 \gg L$, or $a \gg 1$, let us write the integral in the form

$$T = \frac{L}{\sqrt{gh_0}}\,\frac{\sqrt{2}}{\pi}\int_0^{(\pi/2)-\varepsilon} d\xi\sqrt{\frac{1 + a^2\cos^2\xi}{\sin\xi}} + \frac{L}{\sqrt{gh_0}}\,\frac{\sqrt{2}}{\pi}\int_{(\pi/2)-\varepsilon}^{\pi/2} d\xi\sqrt{\frac{1 + a^2\cos^2\xi}{\sin\xi}}$$

We can neglect 1 compared to $a^2 \cos^2 \xi$ for the region from 0 to $(\pi/2) - \varepsilon$, with $1/a < \varepsilon \ll 1$. Then we have

$$
T \approx \frac{L}{\sqrt{gh_0}} \frac{\sqrt{2}}{\pi} \int_0^{(\pi/2)-\varepsilon} d\xi \frac{a|\cos\xi|}{\sqrt{\sin\xi}} \approx \frac{L}{\sqrt{gh_0}} \frac{\sqrt{2}}{\pi} a \int_0^{(\pi/2)-\varepsilon} \frac{d(\sin\xi)}{\sqrt{\sin\xi}}
$$

$$
= \frac{L}{\sqrt{gh_0}} \frac{\sqrt{2}}{\pi} a \cdot 2\sqrt{\sin\xi} \Big|_0^{(\pi/2)-\varepsilon} \approx \sqrt{\frac{2h_0}{g}}
$$

This corresponds to free fall from the height h_0, where $gT^2/2 = h_0$.

1.9 Pulling Strings (MIT)

In order to keep the mass traveling in a circular orbit of radius r, you must apply a force F equal to the mass times its centripetal acceleration v^2/r

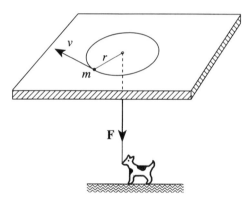

Figure   S.1.9

(see Figure S.1.9). Pulling on the rope exerts no torque on the rotating mass, so the angular momentum $l = mvr$ is conserved. Therefore

$$
F = \frac{mv^2}{r} = \frac{l^2}{mr^3}
$$

Then the work W necessary to move the mass from its initial orbit of radius R to its final orbit of radius $R/2$ is

$$
W = -\int_R^{R/2} F \cdot dr = \frac{l^2}{2mr^2}\Big|_R^{R/2} = \frac{l^2}{2m}\left(\frac{4}{R^2} - \frac{1}{R^2}\right) = \frac{3l^2}{2mR^2}
$$

Solving in terms of E_0 :

$$W = \frac{3m^2 v^2 R^2}{2mR^2} = \frac{3mv^2}{2} = 3E_0$$

1.10 Thru-Earth Train (Stony Brook, Boston (a), Wisconsin-Madison (a))

a) The radial force acting on a particle at any given point inside the Earth depends only on the mass of the sphere whose radius is at that point (see

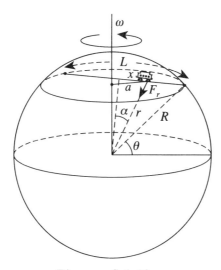

Figure **S.1.10 a**

Figure S.1.10a):

$$F_r = G \frac{mM(r)}{r^2} \qquad M(r) = \frac{4}{3}\pi \rho r^3$$

where m is the mass of the car, so

$$F_r = \frac{4}{3}\pi \rho Gmr$$

The accelerating force will be

$$F_x = F_r \sin \alpha = \frac{4}{3}\pi \rho Gmr \sin \alpha = \frac{4}{3}\pi \rho Gmx$$

so we have

$$m\ddot{x} + \frac{4}{3}\pi\rho Gmx = 0$$

On the surface of the Earth

$$g = \frac{4}{3}\pi\rho GR$$

resulting in

$$\ddot{x} + \frac{g}{R}x = 0$$

which describes oscillatory motion of frequency $\Omega = \sqrt{g/R}$. Half of the period of this oscillation is the time t for the train to get to San Francisco

$$t = \pi\sqrt{\frac{R}{g}} \approx 2540 \text{ s} \approx 42 \text{ minutes}$$

b) If there is friction proportional to the square of velocity we have an additional term

$$\ddot{x} + \frac{\alpha}{m}\dot{x}\,|\dot{x}| + \frac{g}{R}x = 0$$

where α is the proportionality coefficient for the friction force. Using $p = m\dot{x}$ we obtain

$$\dot{p} + \frac{\alpha}{m^2}p\,|p| + \frac{mg}{R}x = 0$$

$$\dot{x} = \frac{p}{m}$$

or

$$\frac{dp}{dx} = -\frac{1}{p}\left(\frac{\alpha p\,|p|}{m} + \frac{m^2 g}{R}x\right)$$

(see Figure S.1.10b).

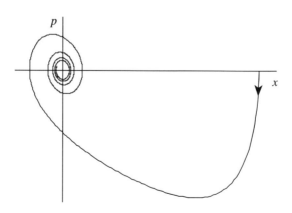

Figure S.1.10 b

c) The acceleration due to the centrifugal force is given by the formula

$$|\mathbf{a}_c| = \frac{|\mathbf{F}_c|}{m} = |-\boldsymbol{\omega} \times \boldsymbol{\omega} \times \mathbf{r}| = \omega^2 a$$

where a is defined in Figure S.1.10a, and $\omega = 2\pi/T$ is the angular frequency of the Earth's rotation. So the maximum centrifugal acceleration $|\mathbf{a}_c|_{max}$ is on the surface of the Earth

$$|\mathbf{a}_c|_{max} = \omega^2 R \cos\theta$$

For New York $\theta \approx 40°$, yielding

$$|\mathbf{a}_c|_{max} = \left(4\pi^2/T^2\right) R \cos\theta \approx 2.6 \text{ cm/s}^2.$$

So the centrifugal force is only about 0.3% of the gravitational force. The acceleration due to the Coriolis force is

$$|\mathbf{a}_{Cor}|_{max} = |-2\boldsymbol{\omega} \times \mathbf{v}_{max}| = 2\omega |\mathbf{v}_{max}|$$

From (a):

$$x = x_0 \cos\Omega t$$

$$x_0 = R\sin\phi$$

where $\phi = L/2R = 2500/6400 \approx 0.39$ rad. So,

$$v = \dot{x} = -R\Omega \sin\phi \sin\Omega t$$

and $|\mathbf{a}_{Cor}|_{max} = 2\omega\Omega R \sin\phi$, where Ω is the frequency of oscillation found in (a): $\Omega = \sqrt{g/R}$, and so $|\mathbf{a}_{Cor}|_{max} = 2\omega\sqrt{gR}\sin\phi \approx 45 \text{ cm/s}^2$. Hence, the Coriolis force is about 5% of the gravitational force.

1.11 String Oscillations (Moscow Phys-Tech)

We have assumed that the dependence is of the form

$$f \propto L^\alpha T^\beta \rho^\gamma$$

To find α, β, γ, we use dimensional analysis, i.e., assume that the dimensions are equivalent on both sides of our proportion:

$$[f] \propto [L]^\alpha [T]^\beta [\rho]^\gamma$$

In MKS units, we have

$$S^{-1} = (M)^\alpha \left(\frac{Kg \cdot M}{S^2}\right)^\beta \left(\frac{Kg}{M}\right)^\gamma$$

or

$$S^{-1} = M^{\alpha + \beta - \gamma} \cdot Kg^{\beta + \gamma} \cdot S^{-2\beta}$$

This is true if the following equations hold:

$$\begin{cases} -2\beta = -1 & \beta = 1/2 \\[2mm] \beta + \gamma = 0 & \Rightarrow \quad \gamma = -1/2 \\[2mm] \alpha + \beta - \gamma = 0 & \alpha = -1 \end{cases}$$

So

$$f \propto L^{-1} T^{1/2} \rho^{-1/2} = \frac{1}{L}\sqrt{\frac{T}{\rho}}$$

1.12 Hovering Helicopter (Moscow Phys-Tech)

The simplest model for a helicopter rotor is a disk of area A inducing a uniform flow of air with velocity v close to the rotor and w far downstream. For an estimate, this model is adequate (see for instance Johnson, *Helicopter Theory*). We disregard any energy loss due to turbulence and consider air to be an incompressible fluid. The rate of mass flowing through the area A of the rotor will be

$$\dot{m} = \rho A v \qquad\qquad\qquad (S.1.12.1)$$

The thrust T is equal to the momentum change per unit time of the air from velocity 0 to w :

$$T = \dot{m} w \qquad\qquad\qquad (S.1.12.2)$$

The power P is equal to the energy change of the same amount of air per unit time (seconds):

$$P = T \cdot v = \frac{1}{2}\dot{m} w^2 \qquad\qquad\qquad (S.1.12.3)$$

From (S.1.12.2) and (S.1.12.3) we have $w = 2v$, and substituting $\dot{m}$ and w from (S.1.12.1) into (S.1.12.3), we can get

$$P = Tv = \frac{1}{2}\rho A v \left(2v\right)^2 = 2\rho A v^3 \qquad\qquad\qquad (S.1.12.4)$$

and therefore
$$T = 2\rho A v^2 \qquad (S.1.12.5)$$
If the helicopter is hovering, it means that its weight W is equal to the thrust T in our ideal case. Then
$$W = 2\rho A v^2 \qquad v = \sqrt{W/2\rho A} \qquad (S.1.12.6)$$
The resulting power is
$$P = Tv = W\sqrt{W/2\rho A} \qquad (S.1.12.7)$$

Since the weight W is proportional to the volume L^3 of the helicopter, and the area A is, of course, proportional to L^2, we find
$$P \propto L^3 \sqrt{L^3/L^2} = L^{7/2} \qquad (S.1.12.8)$$
So, for a model of the helicopter
$$p = \left(\frac{l}{L}\right)^{7/2} P \qquad (S.1.12.9)$$

where p and l are the power and size of the model, respectively. For a 1/10th size model:
$$p = \left(\frac{1}{10}\right)^{7/2} P = \left(\frac{1}{10}\right)^{7/2} \cdot 100\,\text{hp} \approx 3.16 \cdot 10^{-2}\,\text{hp} \approx 24\,\text{W}$$

1.13   Astronaut Tether (Moscow Phys-Tech, Michigan)

The spaceship moves under the influence of the Earth's gravity, given by

$$F = G\frac{MM_E}{R_1^2} \qquad (S.1.13.1)$$

where M is the mass of the spaceship, M_E is the Earth's mass, R_1 is the distance between the center of the Earth and the ship, and G is the gravitational constant (see Figure S.1.13). We may write for the spaceship

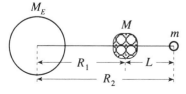

Figure S.1.13

$$G\frac{MM_E}{R_1^2} - T = M\Omega^2 R_1 \qquad (S.1.13.2)$$

where Ω is the angular velocity of the ship and T is the tension of the communication cable. Similarly, for the astronaut,

$$G\frac{mM_E}{R_2^2} + T = m\Omega^2 R_2 \qquad (S.1.13.3)$$

where m is the mass of the astronaut, R_2 is the radius of the orbit of the astronaut, and Ω is the angular velocity of his rotation about the Earth. Obviously, we can write (S.1.13.2) and (S.1.13.3) with the same angular velocity only for the specific case where the spaceship and the astronaut fall on a ray from the Earth's center. Equating Ω from (S.1.13.2) and (S.1.13.3), we obtain

$$\frac{1}{MR_1}\left(G\frac{MM_E}{R_1^2} - T\right) = \frac{1}{mR_2}\left(G\frac{mM_E}{R_2^2} + T\right) \qquad (S.1.13.4)$$

From equation (S.1.13.4), we can easily find the tension T :

$$T = \frac{Mm}{MR_1 + mR_2}\left(\frac{R_2^3 - R_1^3}{R_1^2 R_2^2}\right)M_E G$$

Using $R_1 \approx R_2 \approx R$, $R_2^3 - R_1^3 = (R_2 - R_1)\left(R_2^2 + R_1 R_2 + R_1^2\right) \approx 3R^2 L$, we can rewrite T in the form:

$$T = 3\frac{Mm}{M+m}L\frac{M_E G}{R^3} = 3\frac{L}{R}\frac{Mm}{M+m}g,$$

where g is the acceleration on the surface of the Earth. Also, since $M \gg m$, we can write an even simpler formula as an estimate:

$$T = 3\frac{Lmg}{R} = 3\cdot\frac{100\cdot 110\cdot 9.8}{6400\cdot 10^3} \approx 0.05\,\mathrm{N}$$

Hence, the wire would withstand the tension of holding the hapless astronaut in tow.

1.14   Spiral Orbit (MIT)

The solution may be obtained most quickly by employing the differential equation for the orbit (see Goldstein, *Classical Mechanics*, §3-5), whereby the time dependence is eliminated from the equation of motion. The derivation proceeds from the definition of angular momentum l, which is conserved in central force motion, and consists of the substitution of $\left(l/mr^2\right)(d/d\theta)$

for d/dt. In its final form, the equation reads as follows:

$$\frac{d^2 u}{d\theta^2} + u = -\frac{m}{l^2}\frac{d}{du}V\left(\frac{1}{u}\right) \qquad u \equiv \frac{1}{r} \qquad \text{(S.1.14.1)}$$

We now substitute the proposed potential and orbit equation into (S.1.14.1):

$$u = \frac{1}{c\theta^2} \qquad\qquad \frac{d^2 u}{d\theta^2} = \frac{6}{c\theta^4}$$

$$\frac{dV}{du} = -p\alpha u^{-(p+1)} - q\beta u^{-(q+1)}$$

yielding

$$\frac{6}{c\theta^4} + \frac{1}{c\theta^2} = \frac{m}{l^2}\left(p\alpha c^{p+1}\theta^{2(p+1)} + q\beta c^{q+1}\theta^{2(q+1)}\right) \qquad \text{(S.1.14.2)}$$

Identifying powers of θ on the two sides of (S.1.14.2) gives

$$-4 = 2(p+1)$$
$$-2 = 2(q+1)$$

and therefore

$$p = -3$$
$$q = -2$$

1.15   Central Force with Origin on Circle (MIT, Michigan State)

The differential equation of the orbit comes to the rescue here as in problem 1.14. From Figure S.1.15, we see that the orbit equation is

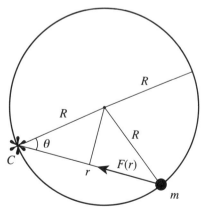

Figure S.1.15

$$r = 2R\cos\theta \qquad (S.1.15.1)$$

so

$$u = \frac{1}{r} = \frac{1}{2R}\frac{1}{\cos\theta} \qquad (S.1.15.2)$$

and

$$\frac{du}{d\theta} = \frac{1}{2R}\frac{\sin\theta}{\cos^2\theta} \qquad \frac{d^2u}{d\theta^2} = \frac{1}{2R}\frac{1+\sin^2\theta}{\cos^3\theta} \qquad (S.1.15.3)$$

Applying the differential equation of the orbit and substituting (S.1.15.2) and (S.1.15.3) into it, we obtain

$$\frac{d^2u}{d\theta^2} + u = -\frac{m}{l^2}\frac{d}{du}V\left(\frac{1}{u}\right) = \frac{1}{2R}\left(\frac{1+\sin^2\theta}{\cos^3\theta} + \frac{1}{\cos\theta}\right)$$

$$= \frac{1}{R\cos^3\theta} = 8R^2u^3 \qquad (S.1.15.4)$$

We are required to find

$$F\left(r\right) = -\frac{dV}{dr} = -\frac{du}{dr}\frac{dV}{du} = \frac{1}{r^2}\frac{dV}{du} \qquad (S.1.15.5)$$

so

$$\frac{8R^2}{r^3} = -\frac{m}{l^2}r^2F\left(r\right) \qquad (S.1.15.6)$$

and finally

$$F\left(r\right) = -\frac{8R^2l^2}{m}\frac{1}{r^5}$$

1.16   Central Force Orbit (Princeton)

a) We see that the orbit describes a cardioid as shown in Figure S.1.16. Invoking the orbit equation yet again (see Problem 1.14), we may find the

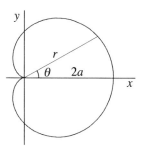

Figure **S.1.16**

force:

$$\frac{l^2 u^2}{m} \left(\frac{d^2 u}{d\theta^2} + u \right) = -f \left(\frac{1}{u} \right) \tag{S.1.16.1}$$

where $u = 1/r$. Calculating the derivatives of u:

$$u = \frac{1}{a \left(1 + \cos\theta \right)}$$

$$\frac{du}{d\theta} = \frac{\sin\theta}{a \left(1 + \cos\theta \right)^2}$$

$$\frac{d^2 u}{d\theta^2} = \frac{1}{a} \frac{\left(1 + \cos\theta \right) \cos\theta + 2 \sin^2\theta}{\left(1 + \cos\theta \right)^3}$$

$$= \frac{1}{a} \frac{\left(1 + \cos\theta \right) \cos\theta + 2 \left(1 + \cos\theta \right) \left(1 - \cos\theta \right)}{\left(1 + \cos\theta \right)^3}$$

$$= \frac{1}{a} \frac{\cos\theta + 2 - 2\cos\theta}{\left(1 + \cos\theta \right)^2}$$

$$= \frac{1}{a} \frac{3 - \left(1 + \cos\theta \right)}{\left(1 + \cos\theta \right)^2}$$

$$= \frac{3a}{a^2 \left(1 + \cos\theta \right)^2} - \frac{1}{a \left(1 + \cos\theta \right)}$$

$$= \frac{3a}{r^2} - \frac{1}{r} \tag{S.1.16.2}$$

and substituting into (S.1.16.1), we obtain

$$f(r) = -\frac{3l^2 a}{mr^4}$$

b) (see Landau and Lifshitz, *Mechanics*, §18). The initial impulse to solve for the scattering angle as a function of the impact parameter leads one astray into the realm of elliptic integrals. Instead, realize that the operative word is "capture" and construct the effective potential of the particle, where

$$U_{\text{eff}} = \frac{l^2}{2mr^2} - \frac{A}{r^4}$$

and A is a constant of proportionality, and l is the angular momentum. Those particles whose kinetic energy exceeds $U_{\text{eff}}(r_0)$ will be captured. At

r_0, $dU\left(r_0\right)/dr = 0$, so we obtain

$$-\frac{l^2}{mr_0^3} + \frac{4A}{r_0^5} = 0 \qquad r_0 = \sqrt{\frac{4Am}{l^2}}$$

and

$$U_{\text{eff}}\left(r_0\right) = \frac{l^4}{8Am^2} - \frac{Al^4}{16A^2m^2} = \frac{l^4}{16Am^2}$$

The condition for capture becomes

$$\frac{1}{2}mv_\infty^2 > \frac{l^4}{16Am^2}$$

where $l = mv_\infty b$, and b is the impact parameter. Rearranging, we find that $b < \sqrt[4]{8A/mv_\infty^2}$. The cross section is given by πb^2, so

$$\sigma_{\text{capture}} = 2\pi\sqrt{\frac{2A}{mv_\infty^2}}$$

1.17 Dumbbell Satellite (Maryland, MIT, Michigan State)

Write the Lagrangian in the frame with the origin at the center of the Earth. The potential energy of the satellite is

$$V = -G\frac{mM}{R_+} - G\frac{mM}{R_-}$$

where M is the mass of the Earth, and R_+ and R_- are the distances from the center of the Earth to the two masses (see Figure S.1.17). Using the formula $R_\pm = \sqrt{R^2 + l^2 \pm 2Rl\cos\theta} \approx R\sqrt{1 \pm (2l/R)\cos\theta}$ (where we

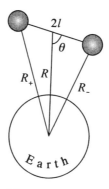

Figure S.1.17

disregard the quadratic term $(l/R)^2$, we can rewrite the potential energy in the form

$$V = -G\frac{mM}{R}\left(\frac{1}{\sqrt{1+(2l/R)\cos\theta}} + \frac{1}{\sqrt{1-(2l/R)\cos\theta}}\right) \quad \text{(S.1.17.1)}$$

Keeping two terms in the expansion of the square root, we obtain

$$V = -G\frac{mM}{R}\left(2 + 2\cdot\frac{3}{8}\left(\frac{2l}{R}\right)^2\cos^2\theta\right) \quad \text{(S.1.17.2)}$$

We can find the minimum of the potential energy now by solving $\partial V/\partial\theta = 0$, which has two solutions:

$$\theta = 0$$

and

$$\theta = \frac{\pi}{2}$$

For the first solution, $\partial^2 V/\partial\theta^2 > 0$, and for the second, $\partial^2 V/\partial\theta^2 < 0$. So, at $\theta = 0$, the potential energy has a minimum, and that determines the orientation of the satellite.

b) For small oscillations

$$\cos^2\theta \approx 1 - \theta^2$$

From (S.1.17.2) we obtain

$$V(\theta) = 3G\frac{mMl^2\theta^2}{R^3} + \text{const} \quad \text{(S.1.17.3)}$$

The kinetic energy of the satellite can be written in the same approximation as a sum of its center of mass energy (which is constant) plus the kinetic energy relative to the center of mass:

$$T = 2\cdot\frac{1}{2}m\left(l\dot\theta\right)^2 + \text{const}$$

So the Lagrangian is

$$\mathcal{L} = T - V = ml^2\dot\theta^2 - 3G\frac{mM}{R^3}l^2\theta^2 + \text{const}$$

The angular velocity ω_0 of the satellite about the Earth may be obtained from the equation for a circular orbit: $m\omega_0^2 R = G\left(mM/R^2\right)$. From the Lagrangian, we arrive at the angular velocity Ω of small angle oscillations of the satellite, where $\Omega^2 = 3G\left(M/R^3\right) = 3\omega_0^2$ and so $\Omega = \sqrt{3}\omega_0$. For further details, see Barger and Olsson, *Classical Mechanics: A Modern Perspective*, §7.3.

1.18 Yukawa Force Orbit (Stony Brook)

The motion can be investigated in terms of the effective potential

$$V_{\text{eff}} = \frac{l^2}{2mr^2} + V(r) \tag{S.1.18.1}$$

where l is the angular momentum of the particle about the origin and

$$F(r) = -\frac{\partial V(r)}{\partial r}$$

The conditions for a stable orbit are

$$\left.\frac{\partial V_{\text{eff}}(r)}{\partial r}\right|_{r=R} = 0 \qquad \left.\frac{\partial^2 V_{\text{eff}}(r)}{\partial r^2}\right|_{r=R} > 0 \tag{S.1.18.2}$$

where $r = R$ is an equilibrium point for the particle in this now one-dimensional problem. The requirement on the second derivative implies that the effective potential is a minimum, i.e., the orbit is stable to small perturbations. Substituting (S.1.18.1) into (S.1.18.2), we obtain

$$\frac{\partial V_{\text{eff}}}{\partial r} = -\frac{l^2}{mr^3} + \frac{\partial V}{\partial r} = \left\{ -\frac{l^2}{mr^3} + \frac{K}{r^2}e^{-r/a} \right\}\bigg|_{r=R} = 0 \tag{S.1.18.3}$$

The second condition of (S.1.18.2) gives

$$\frac{\partial^2 V_{\text{eff}}}{\partial^2 r} = \frac{1}{r^4}\left\{ \frac{3l^2}{m} - 2Kre^{-r/a} - \frac{Kr^2}{a}e^{-r/a} \right\}\bigg|_{r=R} > 0 \tag{S.1.18.4}$$

(S.1.18.3) gives

$$KRe^{-R/a} = \frac{l^2}{m}$$

which, substituted into (S.1.18.4), yields

$$\left.\frac{\partial^2 V_{\text{eff}}}{\partial r^2}\right|_{r=R} = \frac{l^2}{mR^4}\left\{ 1 - \frac{R}{a} \right\} > 0 \tag{S.1.18.5}$$

which implies that, for stability, $a > R$.

b) The equation for small radial oscillations with $\xi = r - R$ is

$$m\ddot{\xi} + \left.\frac{\partial^2 V_{\text{eff}}}{\partial r^2}\right|_{r=R} \cdot \xi = 0 \tag{S.1.18.6}$$

The angular frequency for small oscillations given by (S.1.18.5) and (S.1.18.6) is found from

$$\omega^2 = \frac{1}{m}\frac{\partial^2 V_{\text{eff}}}{\partial r^2}\bigg|_{r=R} = \frac{l^2}{m^2R^4}\left(1 - \frac{R}{a}\right)$$

$$\omega = \frac{l}{mR^2}\sqrt{1 - (R/a)}$$

1.19 Particle Colliding with Reflecting Walls (Stanford)

a) The presence of the perfectly reflecting walls is a smokescreen, obscuring the two-dimensional central force problem (see Figure S.1.19a). In r, θ coordinates, each reflection merely changes $\dot{\theta}$ into $-\dot{\theta}$, which does not affect the energy or the magnitude of the angular momentum, so ignore the walls.

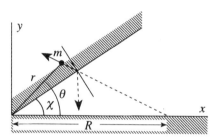

Figure S.1.19a

Write the energy as

$$E = \frac{l^2}{2mr^2} + \frac{1}{2}m\dot{r}^2 - \frac{C}{r^3} \qquad (S.1.19.1)$$

where l is the angular momentum of the particle about the origin. If the particle does not actually hit the origin, at its closest approach to the origin $r = r_{ca}$, $\dot{r} = 0$ (see Figure S.1.19b). Equating the initial energy of the particle with its energy here:

$$E = \frac{1}{2}mV^2 - \frac{C}{R^3} = \frac{l^2}{2mr_{ca}^2} - \frac{C}{r_{ca}^3} \qquad (S.1.19.2)$$

where $l = mV_y R$. Solving (S.1.19.2) for r_{ca} gives the distance of closest approach.

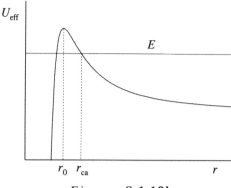

b) Considering the problem in one dimension, we write the effective potential from (S.1.19.1)

$$U_{\text{eff}} = \frac{l^2}{2mr^2} - \frac{C}{r^3} \qquad (\text{S.1.19.3})$$

It has a maximum when $\partial U_{\text{eff}}/\partial r = 0$.

$$-\frac{l^2}{mr_0^3} + \frac{3C}{r_0^4} = 0$$

$$r_0 = \frac{3mC}{l^2} = \frac{3C}{mR^2V_y^2} \qquad (\text{S.1.19.4})$$

Here

$$U_{\text{eff}} = \frac{m^3 R^6 V_y^6}{54C^2} \qquad (\text{S.1.19.5})$$

If the energy of the particle exceeds this value, the particle will greet the
origin and escape to infinity. In addition, if the energy is less than this
value, but the initial position R is less than the value given by (S.1.19.4)

$$R < \left(\frac{3C}{mV_y^2}\right)^{1/3}$$

then the particle will also reach the origin.

c) If, as in (b), the energy exceeds (S.1.19.5), then the particle will escape to infinity. If, on the other hand, the energy is too small, but the
particle starts with $R > r_{ca}$, then the particle will turn around at r_{ca} and
escape also.

1.20   Earth–Comet Encounter (Princeton)

The total energy of the comet is zero since its trajectory is parabolic. In general,

$$E = \frac{m\dot{r}^2}{2} + \frac{l^2}{2mr^2} + U(r) \tag{S.1.20.1}$$

where r is the comet's distance from the Sun, l is its angular momentum, and $U(r)$ is its potential energy (see Figure S.1.20). $U(r) = -GmM_{\text{Sun}}/r$,

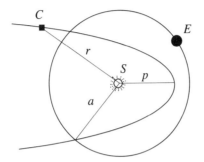

Figure S.1.20

where G is the gravitational constant. Find the total angular momentum defined at the perihelion, where

$$r = r_{\text{min}} \equiv p, \quad \dot{r} = 0$$

$$\frac{l^2}{2mp^2} + U(p) = E = 0$$

Therefore,

$$l = (2m\alpha p)^{1/2} \tag{S.1.20.2}$$

where $\alpha = GmM_{\text{Sun}}$. From (S.1.20.1)

$$\frac{dr}{dt} = \sqrt{\frac{2}{m}\left(\frac{\alpha}{r} - \frac{l^2}{2mr^2}\right)}$$

so the time the comet spends inside the Earth's trajectory is

$$t = 2\int_{r_{\text{min}}}^{r_{\text{max}}} \frac{dr}{\sqrt{\frac{2}{m}\left(\frac{\alpha}{r} - \frac{l^2}{2mr^2}\right)}}$$

But $l^2 = 2m\alpha p$, so from (S.1.20.2)

$$t = 2 \int_{r_{\min}}^{r_{\max}} \frac{dr}{\sqrt{\frac{2}{m}\left(\frac{\alpha}{r} - \frac{\alpha p}{r^2}\right)}} = 2\sqrt{m/2\alpha} \int_p^a \frac{r\,dr}{\sqrt{r - p}} \qquad \text{(S.1.20.3)}$$

where a is the radius of the Earth's orbit. The expression $\int_p^a r\,dr/\sqrt{r - p}$ can be easily integrated by parts, yielding

$$\int_p^a \frac{r\,dr}{\sqrt{r - p}} = \frac{2}{3}a^{3/2}\left(1 + \frac{2p}{a}\right)\sqrt{1 - \frac{p}{a}}$$

Substituting this result back into (S.1.20.3) gives

$$t = \frac{2\sqrt{2}}{3}\sqrt{\frac{ma^3}{\alpha}}\left(1 + \frac{2p}{a}\right)\sqrt{1 - \frac{p}{a}} \qquad \text{(S.1.20.4)}$$

We know that the period of the Earth's revolution about the Sun T_E equals one year, and noting that $T_E = 2\pi\sqrt{ma^3/\alpha}$, we can rewrite (S.1.20.4) in the form

$$t = \frac{\sqrt{2}}{3\pi}T_E\left(1 + \frac{2p}{a}\right)\sqrt{1 - \frac{p}{a}}$$

Denoting $p/a = \xi$, we find the maximum of $F(\xi) = (1 + 2\xi)\sqrt{1 - \xi}$, given that $t = \left(\sqrt{2}/3\pi\right)T_E F(\xi)$:

$$\frac{dF}{d\xi} = 0 \Rightarrow \xi = \frac{1}{2}, \quad F(\xi)\Big|_{1/2} = \sqrt{2}$$

Therefore $t_{\max} = 2T_E/3\pi \approx 77$ days $= 11$ weeks.

1.21   Neutron Scattering (Moscow Phys-Tech)

Consider a neutron colliding with atoms of a certain type. In each collision, neutrons lose a fraction of their kinetic energy; let us calculate this fraction. We will assume that the collision is elastic and central. From energy and momentum conservation,

$$\frac{mv_0^2}{2} = \frac{MV^2}{2} + \frac{mv^2}{2}$$

$$mv_0 = MV + mv$$

where m and M are the masses of the neutron and the atom, respectively;

v_0 and v are the initial and final velocities of the neutron; V is the velocity of the atom after the collision. These equations may be rewritten in the form

$$v_0^2 - v^2 = \frac{M}{m}V^2$$

$$v_0 - v = \frac{M}{m}V$$

Solving for V gives

$$V = \frac{2}{\alpha + 1}v_0$$

letting $\alpha = M/m$. The kinetic energy of the atom after collision is

$$T_A = \frac{MV^2}{2} = \frac{Mv_0^2}{2}\frac{4}{(\alpha + 1)^2} = 4\frac{M}{m}\frac{mv_0^2}{2}\frac{1}{(\alpha + 1)^2}$$

$$= 4\frac{\alpha}{(\alpha + 1)^2}T_0 = 4\frac{T_0}{\alpha + 2 + (1/\alpha)}$$

where T_0 is the initial kinetic energy of the neutron. Obviously, as $\alpha \to 0$ or ∞, $T_A \to 0$. The maximum of T_A as a function of α corresponds to the minimum of $(\alpha + 2 + 1/\alpha)$. So we have $\alpha = 1$ (α positive). Here, $m = M$ and the kinetic energy of the atom as a result of the collision will be a maximum $T_A = T_0$. For hydrogen, $\alpha = M_H/m$ is very close to 1 and this explains why materials with high hydrogen content are so efficient in blocking the neutrons.

1.22 Collision of Mass–Spring System (MIT)

a) The maximum compression of the spring occurs at the moment when the velocities of the two masses m_1 and m_2 become equal (see Figure S.1.22). For this moment we can write conservation of momentum and energy as

$$m_1V_0 = (m_1 + m_2)V' \qquad (S.1.22.1)$$

$$\frac{m_1}{2}V_0^2 = \frac{m_1 + m_2}{2}V'^2 + \frac{kA^2}{2} \qquad (S.1.22.2)$$

where A is the maximum compression of the spring; from (S.1.22.1)

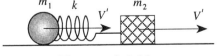

Figure S.1.22

$$V' = \frac{m_1}{m_1 + m_2} V_0$$

and from (S.1.22.2)

$$A = \sqrt{\mu/k}\, V_0$$

where $\mu = m_1 m_2 / (m_1 + m_2)$ is the reduced mass.

b) If, long after the collision, both masses move in the same direction, it means that $m_1 > m_2$, and the spring will not be compressed. So we have

$$\frac{m_1 V_0^2}{2} = \frac{m_1 V_1^2}{2} + \frac{m_2 V_2^2}{2}$$

$$m_1 V_0 = m_1 V_1 + m_2 V_2$$

We can easily find V_1 and V_2 from these equations

$$V_1 = \frac{m_1 - m_2}{m_1 + m_2} V_0 = \frac{1 - \gamma}{1 + \gamma} V_0$$

$$V_2 = \frac{2m_1}{m_1 + m_2} V_0 = \frac{2}{1 + \gamma} V_0$$

where $\gamma = m_2/m_1$.

1.23 Double Collision of Mass–Spring System (Moscow Phys-Tech)

a) Let us call the ball of mass M ball 1, the first ball struck 2, and the third ball 3 (see Figure S.1.23a). After the first collision, ball 1 will move with constant velocity V_1 and balls 2 and 3 will oscillate. For another collision to take place, the coordinates of balls 1 and 2 must coincide at some later moment. First find V_1 after the initial collision, considered to be instantaneous. Then, this problem is no different from the collision of just two balls of masses M and m. If the velocity of the first ball before the collision is V_0 we can find V_1 and V_2 from energy and momentum

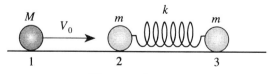

Figure S.1.23a

conservation:

$$MV_0 = MV_1 + mV_2$$

$$\frac{MV_0^2}{2} = \frac{MV_1^2}{2} + \frac{mV_2^2}{2}$$

Again, as in Problem 1.22,

$$V_1 = \frac{M-m}{M+m}V_0 = \frac{1-\gamma}{1+\gamma}V_0 \qquad (S.1.23.3)$$

$$V_2 = \frac{2M}{M+m}V_0 = \frac{2}{1+\gamma}V_0 \qquad (S.1.23.4)$$

$$\gamma \equiv \frac{m}{M}$$

After the collision the first ball will move with constant velocity V_1 and so its position coordinate $x_1 = V_1 t = V_0 t (1-\gamma)/(1+\gamma)$. The center of mass of balls 2 and 3 will also move with constant velocity $V_c = V_2/2$ (since $m_2 = m_3 = m$). Therefore from (S.1.23.4)

$$x_c = \frac{V_0}{1+\gamma} \cdot t$$

Now, in the center of mass frame of balls 2 and 3, the two balls are moving toward one another each with speed $V_2/2$, and they will start to oscillate relative to the center of mass with the following time dependence:

$$x_{2,3}(t) = A \sin \omega t$$

where $\omega = \sqrt{k'/m}$ and k' is the spring constant of half of the spring, $k' \equiv 2k$. From energy conservation, the initial energy of mass 2 in the center of mass frame goes into the energy of spring deformation with an amplitude corresponding to the velocity change from $V_2/2$ to zero:

$$\frac{m(V_2/2)^2}{2} = \frac{mV_0^2}{2(1+\gamma)^2} = \frac{k'A^2}{2}$$

$$A = \sqrt{mV_0^2/k'(1+\gamma)^2} = \frac{V_0}{(1+\gamma)\omega}$$

In the lab frame

$$x_2(t) = x_c(t) + A \sin \omega t = \frac{V_0 t}{1+\gamma} + \frac{V_0}{(1+\gamma)\omega} \sin \omega t$$

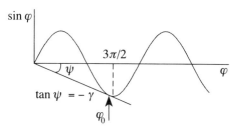

Figure **S.1.23b**

For the second collision to occur, we need $x_1 = x_2$ or

$$\frac{1-\gamma}{1+\gamma} V_0 t = \frac{V_0 t}{1+\gamma}\left(1+\frac{\sin\omega t}{\omega t}\right) \qquad (\text{S.1.23.5})$$

So we have

$$\frac{\sin\omega t}{\omega t} \equiv \frac{\sin\varphi}{\varphi} = -\gamma \qquad (\text{S.1.23.6})$$

The easiest way is to solve (S.1.23.6) graphically (see Figure S.1.23b). For the solution to exist, we have the condition $\gamma \leq \gamma_{\text{max}}$, where

$$\gamma_{\text{max}} \approx \frac{1}{3\pi/2} \approx 0.2$$

at $\varphi_0 \approx 3\pi/2$. The minimum value of the mass $M = m/\gamma_{\text{max}} \approx 10$ kg.

b) The time between collisions is

$$t_0 = \frac{\varphi_0}{\omega} \approx \frac{3\pi}{2}\sqrt{\frac{m}{k'}} = \frac{3\pi}{2}\sqrt{\frac{m}{2k}} \approx 5 \text{ s}$$

1.24   Small Particle in Bowl (Stony Brook)

a) In spherical coordinates, the Lagrangian

$$\mathcal{L} = T - V = \frac{m}{2}\left(\dot{r}^2 + r^2\dot{\varphi}^2\sin^2\theta + r^2\dot{\theta}^2\right) + mgr\cos\theta$$

Since we have the restriction $r = \text{constant} = R$ (see Figure S.1.24),

$$\mathcal{L} = \frac{mR^2}{2}\left(\dot{\varphi}^2\sin^2\theta + \dot{\theta}^2\right) + mgR\cos\theta \qquad (\text{S.1.24.1})$$

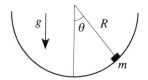

Figure **S.1.24**

b) For the generalized momenta

$$p_\theta = \frac{\partial \mathcal{L}}{\partial \dot\theta} = mR^2\dot\theta$$

(S.1.24.2)

$$p_\varphi = \frac{\partial \mathcal{L}}{\partial \dot\varphi} = mR^2\dot\varphi \sin^2\theta$$

c) Find the Hamiltonian of the motion $\mathcal{H}(p,q,t)$

$$\mathcal{H}(p,q,t) = \sum p_i\dot q_i - \mathcal{L}$$

From (b)

$$\dot\theta = \frac{p_\theta}{mR^2}$$

(S.1.24.3)

$$\dot\varphi = \frac{p_\varphi}{mR^2 \sin^2\theta}$$

(S.1.24.4)

The Hamiltonian then becomes

$$\mathcal{H} = \frac{1}{2mR^2}\left(\frac{p_\varphi^2}{\sin^2\theta} + p_\theta^2\right) - mgR\cos\theta$$

(S.1.24.5)

d) Let us write Hamilton's equations

$$\dot p_i = -\frac{\partial \mathcal{H}}{\partial q_i}$$

$$\dot q_i = \frac{\partial \mathcal{H}}{\partial p_i}$$

$$\dot p_\theta = -\frac{\partial \mathcal{H}}{\partial \theta} = \frac{p_\varphi^2 \cos\theta}{mR^2 \sin^3\theta} - mgR\sin\theta$$

(S.1.24.6)

$$\dot p_\varphi = -\frac{\partial \mathcal{H}}{\partial \varphi} = 0 \qquad p_\varphi = p_{\varphi_0} = \text{const}$$

(S.1.24.7)

e) Differentiate (S.1.24.3) and use (S.1.24.6) and (S.1.24.7)

$$\ddot{\theta} = \frac{\dot{p}_\theta}{mR^2} = \frac{1}{mR^2}\left(\frac{1}{mR^2}\frac{p_{\varphi_0}^2 \cos\theta}{\sin^3\theta} - mgR\sin\theta\right) \quad \text{(S.1.24.8)}$$

f) If $\theta = \theta_0$, $\dot{\theta} = 0$, $\ddot{\theta} = 0$, we have

$$\frac{1}{mR^2}\frac{p_{\varphi_0}^2 \cos\theta_0}{\sin^3\theta_0} - mgR\sin\theta_0 = 0 \quad \text{(S.1.24.9)}$$

$$p_{\varphi_0} = \pm mR\sqrt{gR/\cos\theta_0}\,\sin^2\theta_0$$

$$\text{(S.1.24.10)}$$

$$\dot{\varphi} = \frac{p_{\varphi_0}}{mR^2\sin^2\theta_0} = \pm\sqrt{g/R\cos\theta_0}$$

Here, the particle slides in a circle at a fixed height in the bowl. The different signs correspond to clockwise or counterclockwise motion.

g) If $\theta = \theta_0$, $\dot{\theta} = 0$, $\dot{\varphi} = 0$ at $t = 0$, then we always have

$$p_\varphi = p_{\varphi_0} = 0$$

and so

$$\ddot{\theta} = -\frac{g\sin\theta}{R} \quad \text{(S.1.24.11)}$$

(the equation for a simple pendulum). The energy is conserved and therefore

$$\mathcal{H}(t) = \text{const} = \mathcal{H}(0) \quad \text{(S.1.24.12)}$$

Using (S.1.24.2), (S.1.24.5), and (S.1.24.12), we have

$$\mathcal{H}(t) = \frac{p_\theta^2}{2mR^2} - mgR\cos\theta = -mgR\cos\theta_0$$

and

$$|p_\theta| = mR\sqrt{2gR(\cos\theta - \cos\theta_0)}$$

The maximum velocity corresponds to the maximum $|p_\theta|$ which occurs at $\theta = 0$

$$|p_\theta|_{\max} = mR\sqrt{2gR(1 - \cos\theta_0)}$$

$$|V_{\max}| = R\dot{\theta}_{\max} = \frac{|p_\theta|_{\max}}{mR} = \sqrt{2gR(1 - \cos\theta_0)}$$

Of course, this result can be obtained much more easily from (S.1.24.11) using elementary methods.

1.25   Fast Particle in Bowl (Boston)

Introduce cylindrical coordinates ρ, φ, z, where z is positive down (see Figure S.1.25). We can write the Lagrangian

$$\mathcal{L} = \frac{m}{2} \left(\dot{\rho}^2 + \rho^2 \dot{\varphi}^2 + \dot{z}^2 \right) + mgz \tag{S.1.25.1}$$

From (S.1.25.1), we can see that, as usual, the angular momentum is conserved:

$$m\rho^2 \dot{\varphi} = \text{const} = l_0 = mR^2 \omega$$

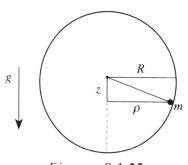

Figure **S.1.25**

Using the constraint $\rho^2 = R^2 - z^2$ which follows from the fact that the particle moves along the spherical surface, we have

$$\dot{\varphi} = \frac{\omega R^2}{\rho^2} = \frac{\omega R^2}{R^2 - z^2} \tag{S.1.25.2}$$

From the same constraint,

$$\dot{\rho} = -\frac{z\dot{z}}{\sqrt{R^2 - z^2}} \tag{S.1.25.3}$$

Energy is of course also conserved:

$$E = \frac{m}{2} \left(\dot{\rho}^2 + \rho^2 \dot{\varphi}^2 + \dot{z}^2 \right) - mgz = E_0 = \frac{m\omega^2 R^2}{2} \tag{S.1.25.4}$$

Substituting $\dot{\varphi}$ and $\dot{\rho}$ from (S.1.25.2) and (S.1.25.3) into (S.1.25.4), we obtain

$$\frac{1}{2}\left(\frac{R^2\dot{z}^2}{R^2-z^2}+\frac{\omega^2 R^4}{R^2-z^2}\right)-gz=\frac{\omega^2 R^2}{2} \qquad \text{(S.1.25.5)}$$

The condition $\omega^2 R \gg g$ leads to $z \ll R$. Therefore we can approximately write

$$\dot{z}^2\left(1+\frac{z^2}{R^2}\right)+\omega^2 R^2\left(1+\frac{z^2}{R^2}\right)-2gz=\omega^2 R^2 \qquad \text{(S.1.25.6)}$$

or

$$\dot{z}^2+\omega^2 z^2-2gz=0$$

which is satisfied by

$$z=\frac{g}{\omega^2}\left(1-\cos\omega t\right)=\frac{2g}{\omega^2}\sin^2\frac{\omega t}{2} \qquad \text{(S.1.25.7)}$$

1.26 Mass Orbiting on Table (Stony Brook, Princeton, Maryland, Michigan)

a) We can write the Lagrangian in terms of the length r of the string on the table and the angle θ (see Figure S.1.26):

$$\mathcal{L}=\frac{1}{2}M\left(\dot{r}^2+r^2\dot{\theta}^2\right)+\frac{1}{2}m\dot{r}^2-mgr$$

The equations of motion are

$$(M+m)\ddot{r}-Mr\dot{\theta}^2+mg=0 \qquad \text{(S.1.26.1)}$$

$$\frac{d}{dt}\left(Mr^2\dot{\theta}\right)=0 \qquad \text{(S.1.26.2)}$$

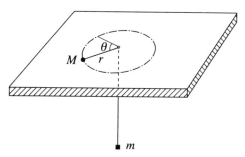

Figure **S.1.26**

From (S.1.26.9) we have angular momentum conservation: $Mr^2\dot\theta = \text{const}$
$= l_0$ so $\dot\theta = l_0/Mr^2$ and

$$\mathcal{L} = \frac{1}{2}(M+m)\dot r^2 + \frac{l_0^2}{2Mr^2} - mgr$$

b) The equilibrium position is defined by taking the derivative of U_{eff} where

$$U_{\text{eff}} = mgr + \frac{l_0^2}{2Mr^2}$$

$$\left.\frac{\partial U_{\text{eff}}}{\partial r}\right|_{r=r_0} = 0$$

$$r_0 = \left(\frac{l_0^2}{gMm}\right)^{1/3}$$

$\partial^2 U_{\text{eff}}/\partial r^2 > 0$, so the orbit is stable with respect to a small perturbation in the radius. The frequency of small oscillations is given by

$$\omega^2 = \frac{1}{M_{\text{eff}}}\left.\left(\frac{\partial^2 U_{\text{eff}}}{\partial r^2}\right)\right|_{r=r_0} = \frac{1}{M+m}\left.\left(\frac{\partial^2 U_{\text{eff}}}{\partial r^2}\right)\right|_{r=r_0}$$

$$= \frac{1}{M+m}\left(\frac{3l_0^2}{Mr_0^4}\right) = \frac{1}{1+(M/m)}\left(\frac{3g}{r_0}\right)$$

1.27 Falling Chimney (Boston, Chicago)

First calculate the motion of the entire chimney of mass m by considering the torque τ about its base B (see Figure S.1.27)

$$\tau = -mg\frac{L}{2}\cos\theta$$

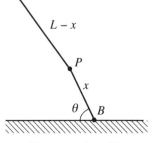

Figure **S.1.27**

The moment of inertia about the base is $I = (1/3) \, mL^2$. The equation of motion for θ is found from

$$\tau = I\ddot{\theta}$$

$$\ddot{\theta} = -\frac{3}{2}\frac{g}{L}\cos\theta$$

The piece of the chimney above the point P rotates in response to the torque τ_{cm} produced by its center of mass about P given by

$$\tau_{cm} = -\frac{m\,(L-x)}{L}\,g\,\frac{(L-x)}{2}\,\cos\theta$$

and the torque $\tau(x)$ produced by the rest of the chimney attached below P, "trying" to convince the piece to rotate at $\ddot{\theta}$

$$\tau\,(x) - \frac{m\,(L-x)}{L}\,g\,\frac{(L-x)}{2}\,\cos\theta = \frac{1}{3}\frac{m\,(L-x)}{L}\,(L-x)^2\,\ddot{\theta}$$

Find $\tau\,(x)$ by substituting for $\cos\theta$ above:

$$\tau\,(x) = \frac{1}{3}\frac{m}{L}\,(L-x)^3\,\ddot{\theta} - \frac{m}{3}\,(L-x)^2\,\ddot{\theta}$$

Find the maximum torque by taking the derivative of $\tau\,(x)$ with respect to x and setting it equal to zero:

$$\frac{d\tau\,(x)}{dx} = -\frac{m}{L}\,(L-x)^2\,\ddot{\theta} + \frac{2m}{3}\,(L-x)\,\ddot{\theta} = 0$$

Either $x = L$, where the torque is zero, or $L - x = (2/3)\,L \Rightarrow x = L/3$, as was to be demonstrated. This problem may also be found in Cronin, Greenberg, Telegdi, *University of Chicago Graduate Problems in Physics* and Routh, *Dynamics of a System of Rigid Bodies*.

1.28   Sliding Ladder (Princeton, Rutgers, Boston)

Let us watch the ladder until it leaves the wall. Forces N_1 and N_2 are normal reactions of the wall and floor, respectively; $W = mg$ is the weight of the ladder; x_c and y_c are the coordinates of the center of mass (see Figure S.1.28). First, find the Lagrangian of the system. The kinetic energy $T = (m/2)\left(\dot{x}_c^2 + \dot{y}_c^2\right) + (1/2)\,I_c\dot{\alpha}^2$. I_c is the moment of inertia relative to the center of mass

$$I_c = \frac{m}{12}\,(2l)^2 = \frac{m}{3}\,l^2$$

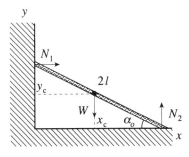

Figure **S.1.28**

$$T = \frac{m}{2}\left(\dot{x}_c^2 + \dot{y}_c^2\right) + \frac{1}{6}ml^2\dot{\alpha}^2 = \frac{2}{3}ml^2\dot{\alpha}^2$$

$$\mathcal{L} = \frac{2}{3}ml^2\dot{\alpha}^2 - mgl\sin\alpha \qquad (S.1.28.1)$$

From Lagrange's equations

$$\frac{4}{3}l\ddot{\alpha} = -g\cos\alpha \qquad (S.1.28.2)$$

In addition, from energy conservation

$$(2/3)\,ml^2\dot{\alpha}^2 = mgl\,(\sin\alpha_0 - \sin\alpha) \qquad (S.1.28.3)$$

We will assume that the ladder loses contact with the wall before it does so with the floor. (This has to be checked). Up until the ladder slides away from the wall, there are constraints of the form

$$x_c = l\cos\alpha \qquad (S.1.28.4)$$

$$y_c = l\sin\alpha \qquad (S.1.28.5)$$

Since N_1 is the only force acting in the x direction, $N_1 = m\ddot{x}_c$. When the ladder loses contact with the wall, $N_1 = 0$. Differentiating (S.1.28.4) twice gives

$$\ddot{x}_c = -l\dot{\alpha}^2\cos\alpha - l\ddot{\alpha}\sin\alpha = 0 \qquad (S.1.28.6)$$

From (S.1.28.6) $\ddot{\alpha} = -\dot{\alpha}^2\cot\alpha$, and substituting it into (S.1.28.2), we have for the angle the ladder leaves the wall

$$\frac{4}{3}l\dot{\alpha}^2 = g\sin\alpha \qquad (S.1.28.7)$$

From (S.1.28.3) and (S.1.28.7), we obtain

$$\frac{1}{2}\sin\alpha = \sin\alpha_0 - \sin\alpha$$

$$\sin\alpha = \frac{2}{3}\sin\alpha_0 \qquad \text{(S.1.28.8)}$$

We have assumed that the ladder loses contact with the wall first. Let us check this assumption. It implies that $N_2 > 0$ at all times before the ladder leaves the wall

$$N_2 = mg + m\ddot{y}_c \qquad \text{(S.1.28.9)}$$

From (S.1.28.5) and (S.1.28.6), we have

$$\ddot{y}_c = -l\dot{\alpha}^2\sin\alpha + l\ddot{\alpha}\cos\alpha = -\frac{l\dot{\alpha}^2}{\sin\alpha} \qquad \text{(S.1.28.10)}$$

Therefore

$$N_2 = mg - \frac{ml\dot{\alpha}^2}{\sin\alpha} \qquad \text{(S.1.28.11)}$$

At the time the ladder leaves the wall

$$N_2 = mg - \frac{ml\dot{\alpha}^2}{\sin\alpha} = mg - \frac{3}{4}mg = \frac{1}{4}mg > 0 \qquad (g > 0)$$

On the other hand, N_2 is monotonically decreasing while α is decreasing (see (S.1.28.3)). So, our assumption was right and indeed $\alpha = \sin^{-1}\left(\frac{2}{3}\sin\alpha_0\right)$.

1.29   Unwinding String (MIT, Maryland (a,b), Chicago (a,b))

a) You can write a Lagrangian using Cartesian coordinates x, y and express it as a function of $\theta, \dot{\theta}$ (see Figure S.1.29a). However, if you notice that the length of the unwound string is $R\theta$, and it unwinds with angular velocity $\dot{\theta}$, you can immediately write a Lagrangian, which is just a kinetic energy in this case

$$\mathcal{L} = \frac{1}{2}mR^2\theta^2\dot{\theta}^2$$

The equation of motion will be

$$\frac{d}{dt}\left(\theta^2\dot{\theta}\right) - \dot{\theta}^2\theta = 0$$

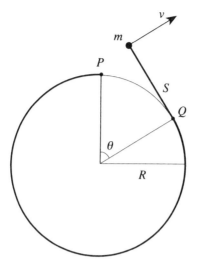

Figure **S.1.29a**

or

$$\theta^2\ddot{\theta} + \dot{\theta}^2\theta = 0$$

or for $\theta \neq 0$,

$$\frac{d}{dt}\left(\dot{\theta}\theta\right) = \frac{d^2}{dt^2}\left(\frac{\theta^2}{2}\right) = 0$$

whose solution is $\theta^2 = 2At + B$. From the initial condition $\theta(0) = 0$ and

$$\mathcal{L}_0 = \frac{mv_0^2}{2}$$

i.e., $\theta(0)\dot{\theta}(0) = v_0/R$, we conclude that $B = 0$ and $A = v_0/R$, so

$$\theta = \sqrt{2\left(v_0/R\right)t}$$

b) The angular momentum l about the center of the cylinder is given by

$$\mathbf{l} = m\mathbf{r} \times \mathbf{v} = mlv_0\hat{\mathbf{e}}_z = mR\theta v_0\hat{\mathbf{e}}_z = m\sqrt{2v_0^3Rt}\,\hat{\mathbf{e}}_z$$

The angular momentum is not conserved, since there is a torque from the cord connected to the cylinder; the energy, on the other hand, is conserved, because the force on the mass m is always perpendicular to its velocity:

$$E = \mathcal{L} = \frac{m}{2}R^2\dot{\theta}^2\theta^2 = \frac{mv_0^2}{2}$$

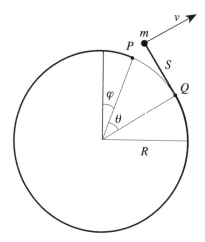

c) Again we can employ cartesian coordinates (see Figure S.1.29b), but if we use the fact that the cord is always perpendicular to the radius of the cylinder at the point of contact we can write the Lagrangian in the form

$$\mathcal{L} = \frac{1}{2} m R^2 \dot{\varphi}^2 + \frac{1}{2} m R^2 \theta^2 \left(\dot{\theta} + \dot{\varphi} \right)^2 + \frac{1}{2} M R^2 \dot{\varphi}^2$$

$$= \frac{1}{2} \left(m + M \right) R^2 \dot{\varphi}^2 + \frac{1}{2} m R^2 \theta^2 \left(\dot{\theta} + \dot{\varphi} \right)^2$$

From this equation, we can immediately obtain the integrals of motion. An angular momentum which is now conserved is

$$P_\varphi = \frac{\partial \mathcal{L}}{\partial \dot{\varphi}} \qquad (\text{S.1.29.1})$$

Initially, it is equal to zero, and since the initial impulse does not give the system any angular momentum, it will be zero for all time. So, we have from (S.1.29.1)

$$(m + M) \dot{\varphi} + m \theta^2 \left(\dot{\theta} + \dot{\varphi} \right) = 0 \qquad (\text{S.1.29.2})$$

The energy is also conserved:

$$E = \frac{1}{2} \left(m + M \right) R^2 \dot{\varphi}^2 + \frac{1}{2} m R^2 \theta^2 \left(\dot{\theta} + \dot{\varphi} \right)^2 = \frac{m v_0^2}{2} \qquad (\text{S.1.29.3})$$

d) From equation (S.1.29.2) we can express $\dot{\varphi}$ in terms of θ and $\dot{\theta}$:

$$\dot{\varphi} = - \frac{\dot{\theta} \theta^2}{\theta^2 + 1 + (M/m)} \qquad (\text{S.1.29.4})$$

Substituting (S.1.29.4) into (S.1.29.3) we obtain

$$\left(1 + \frac{M}{m}\right) \frac{\dot{\theta}^2 \theta^4}{[\theta^2 + 1 + (M/m)]^2} + \theta^2 \left(\dot{\theta} - \frac{\dot{\theta}\theta^2}{\theta^2 + 1 + (M/m)}\right)^2 = \frac{v_0^2}{R^2}$$

After some algebra, letting $\alpha \equiv 1 + (M/m)$ and $\beta \equiv \left(v_0^2/R^2\right)$ we have

$$\frac{\dot{\theta}^2 \theta^2}{\theta^2 + \alpha} = \frac{\beta}{\alpha}$$

Integrating this equation results in

$$\sqrt{\theta^2 + \alpha} = \sqrt{(\beta/\alpha)}\, t + C$$

Therefore:

$$\theta^2 = \frac{\beta}{\alpha} t^2 + 2\sqrt{(\beta/\alpha)}\, Ct + C^2 - \alpha$$

From the initial condition $\theta(0) = 0$, we find that $C^2 = \alpha$, so

$$\theta = \sqrt{(\beta/\alpha)\, t^2 + 2\sqrt{\beta}\, t}$$

Substituting back α and β, we have

$$\theta = \sqrt{\frac{v_0^2 t^2}{R^2 (1 + (M/m))} + \frac{2v_0 t}{R}}$$

For $M/m \to \infty$ (i.e., a fixed cylinder), this result reduces to that obtained in (a). It is obvious from angular momentum conservation that the cylinder would spin in the opposite direction from that of the unwinding cord. Indeed, from (S.1.29.4) we see that if $\dot{\theta} > 0$, then $\dot{\varphi} < 0$. Parts (a) and (b) of this problem were published in Cronin, Greenberg, Telegdi, *University of Chicago Graduate Problems in Physics*.

1.30 Six Uniform Rods (Stony Brook)

This problem, in general (after some arbitrary time t) is rather difficult. However we can use two important simplifications at $t = 0$. First, there is sixfold symmetry, which means that the positions of the center of mass of each rod can be described by just one angle θ and, of course, the length of a rod, which we will denote as $2a$. The other consideration is that even after the blow, the system will keep symmetry relative to the y-axis (because the blow is at midpoint of the first rod). That means that not only at $t = 0$, but also at later times, there will be no rotation of the system, and its angular momentum is zero. We choose the coordinate system as shown in Figure S.1.30. Now the velocity of the center of mass (midpoint P) of the first

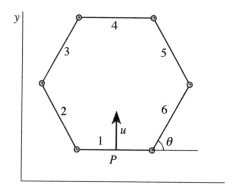

Figure S.1.30

rod is $u = \dot{y}$. The velocity of the midpoint of the opposite one (rod 4) is $\dot{y} + 4a\dot{\theta}\cos\theta$ since the coordinates of its center of mass are $(x, y + 4a\sin\theta)$. We may try to use the condition $P_\theta = 0$ to find a relation between $\dot{\theta}$ and $\dot{y}$. At $t = 0$, $\theta = \pi/3$ for a regular hexagon and

$$P_\theta = \left.\frac{\partial \mathcal{L}}{\partial \dot{\theta}}\right|_{\theta=\pi/3} = \left.\frac{\partial T}{\partial \dot{\theta}}\right|_{\theta=\pi/3}$$

where $\mathcal{L}$ is the Lagrangian of the system, which in this case is equal to its kinetic energy T

$$T = \sum_{i=1}^{6} T_i$$

where $i = 1 \ldots 6$ are subsequent rods (see Figure S.1.30). The kinetic energy of each rod consists of its energy relative to the center of mass plus the energy of its center of mass. Let us say that the mass of the rod is m. The energy relative to the center of mass will be the same for rods $i = 2, 3, 5, 6$ and equal to $T_i = (1/2) I_{\mathrm{cm}}\dot{\theta}^2$, where I_{cm} is the moment of inertia relative to the center of mass. We have already used I_{cm} of a uniform rod in Problem 1.28. To calculate it, we can either integrate (which is very simple in this case) or use a more general approach, which can be applied in other problems with bodies possessing certain symmetries. In this case, we know that the moment of inertia of the rod is $\gamma m L^2$, where L is the length of the rod and γ is some numerical factor. Now move to the edge of the rod. The moment of inertia I_e is

$$I_e = I_{\mathrm{cm}} + m \left(\frac{L}{2}\right)^2$$

where $L/2$ is the distance between the edge and the center of mass. Here we have applied the parallel axis theorem. On the other hand, this I_e is nothing but half of the moment of inertia relative to the center of mass of a rod that is twice as long. So, we can write

$$\frac{1}{2}\gamma\,(2m)\,(2L)^2 = \gamma m L^2 + m\left(\frac{L}{2}\right)^2$$

$$4\gamma = \gamma + \frac{1}{4}$$

$$\gamma = \frac{1}{12}$$

The moment of inertia of the rod is then $I = (1/12)\,mL^2$. In our case, $L = 2a$, so $I = (1/3)\,ma^2$. We then arrive at

$$T_{\text{rel.cm.}} = \frac{1}{2}I\dot{\theta}^2 = \frac{1}{6}ma^2\dot{\theta}^2$$

Now calculate the center of mass energy of each rod $T_i = (m/2)\left(\dot{x}_i^2 + \dot{y}_i^2\right)$:

$$T_1 = \frac{m}{2}\dot{y}^2$$

$$T_2 = T_6 = \frac{m}{2}\left[\left(\dot{y} + a\dot{\theta}\cos\theta\right)^2 + \left(-a\dot{\theta}\sin\theta\right)^2\right]$$

$$T_3 = T_5 = \frac{m}{2}\left[\left(\dot{y} + 3a\dot{\theta}\cos\theta\right)^2 + \left(-a\dot{\theta}\sin\theta\right)^2\right]$$

$$T_4 = \frac{m}{2}\left(\dot{y} + 4a\dot{\theta}\cos\theta\right)^2$$

So the total kinetic energy is

$$T = T_1 + 2T_2 + 2T_3 + T_4 + 4T_{\text{rel.cm.}}$$

$$= 3m\dot{y}^2 + 12ma\dot{\theta}\dot{y}\cos\theta + 8ma^2\dot{\theta}^2\left(\frac{1}{3} + 2\cos^2\theta\right)$$

Now we can calculate P_θ at $\theta = \pi/3$:

$$P_\theta = \left.\frac{\partial T}{\partial \dot{\theta}}\right|_{\theta=\pi/3} = 12ma\dot{y}\cos\theta + 16ma^2\dot{\theta}\left(\frac{1}{3} + 2\cos^2\theta\right)$$

$$= 6ma\dot{y} + \frac{40}{3}ma^2\dot{\theta} = 0$$

So
$$\dot{\theta} = -\frac{9}{20}\frac{\dot{y}}{a}$$

Now recalling that $\dot{y} = u$, we get for the velocity of rod 4

$$v = \dot{y} + 4a\dot{\theta}\cos\theta = \dot{y} + 2a\dot{\theta} = \dot{y} - \frac{9}{10}\dot{y} = \frac{1}{10}\dot{y} = \frac{1}{10}u$$

1.31 Period as Function of Energy (MIT)

Energy is conserved for a position dependent potential, so we may write $E = (1/2)\,m\dot{x}^2 + A|x|^n$. The time for a particle to travel between two turning points of its motion τ_1 and τ_2 (where its kinetic energy is zero) is given by

$$\tau_{12} = \tau_2 - \tau_1 = \int_{\tau_1}^{\tau_2} dt = \int_{x_1}^{x_2} \frac{dx}{\sqrt{(2/m)(E - A|x|^n)}}$$

$$= \int_{-(E/A)^{(1/n)}}^{(E/A)^{(1/n)}} \frac{dx}{\sqrt{(2/m)(E - A|x|^n)}} = 2 \int_{0}^{(E/A)^{(1/n)}} \frac{dx}{\sqrt{(2/m)(E - A|x|^n)}}$$

$$= \sqrt{\frac{2m}{E}} \int_{0}^{(E/A)^{(1/n)}} \frac{dx}{\sqrt{1 - \left[(A/E)^{1/n}x\right]^n}} \qquad \text{(S.1.31.1)}$$

Let $u = (A/E)^{1/n}\,x$. (S.1.31.1) then becomes

$$\tau_{12} = \sqrt{2m}\,E^{\left(\frac{1}{n} - \frac{1}{2}\right)} A^{-(1/n)} \int_{0}^{1} \frac{du}{\sqrt{1 - u^n}} \qquad \text{(S.1.31.2)}$$

The period T is twice the time to go between points 1 and 2, $T = 2\tau_{12}$. So for the energy dependence of the period, we have

$$T \propto E^{\left(\frac{1}{n} - \frac{1}{2}\right)} \qquad \text{(S.1.31.3)}$$

For a harmonic oscillator $n = 2$, and $T = 2\pi/\sqrt{k/m}$, independent of E, as (S.1.31.3) confirms (see Landau and Lifshitz, *Mechanics*, § 11).

1.32   Rotating Pendulum (Princeton, Moscow Phys-Tech)

We may compute the Lagrangian by picking two appropriate orthogonal coordinates θ and φ, where $\dot{\varphi}$ equals a constant ω (see Figure S.1.32).

$$T = \frac{1}{2} m \left(l \sin \theta \right)^2 \omega^2 + \frac{1}{2} m \left(l \dot{\theta} \right)^2$$

$$U = -mgl \cos \theta$$

$$\mathcal{L} = \frac{1}{2} m l^2 \dot{\theta}^2 - \left(-mgl \cos \theta - \frac{1}{2} m l^2 \omega^2 \sin^2 \theta \right)$$

where we consider $U_{\text{eff}} = -\left(1/2 \right) m l^2 \omega^2 \sin^2 \theta - mgl \cos \theta$.

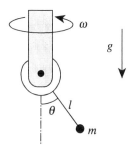

Figure S.1.32

a) Employing the usual Lagrange equations

$$\frac{d}{dt} \frac{\partial \mathcal{L}}{\partial \dot{\theta}} = \frac{\partial \mathcal{L}}{\partial \theta}$$

we have

$$ml^2 \ddot{\theta} - ml^2 \omega^2 \sin \theta \cos \theta + mgl \sin \theta = 0 \qquad \text{(S.1.32.1)}$$

b) (S.1.32.1) has stationary points where $\partial U_{\text{eff}} / \partial \theta = 0$.

$$\sin \theta = 0 \qquad\qquad \theta_1 = 0, \ \theta_2 = \pi$$

$$\cos \theta = g/\omega^2 l \qquad \theta_3 = \cos^{-1} \frac{g}{\omega^2 l} \qquad\qquad g/l \leq \omega^2$$

To check these points for stability, take the second derivative

$$\frac{\partial^2 U_{\text{eff}}}{\partial \theta^2} = -ml^2 \omega^2 \cos^2 \theta + ml^2 \omega^2 \sin^2 \theta + mgl \cos \theta \qquad \text{(S.1.32.2)}$$

At $\theta = 0$ (S.1.32.2) becomes

$$\left.\frac{\partial^2 U_{\text{eff}}}{\partial \theta^2}\right|_{\theta=0} = -ml^2\omega^2 + mgl = ml\left(g - \omega^2 l\right) = \begin{cases} > 0, & \omega^2 < g/l \\ < 0, & \omega^2 > g/l \end{cases}$$

So at angular velocities $\omega_c < \sqrt{g/l}$, the potential energy has a minimum and the $\theta - 0$ equilibrium point is stable. However at $\omega_c > \sqrt{g/l}$, this point is no longer stable. At $\theta = \pi$:

$$\left.\frac{\partial^2 U_{\text{eff}}}{\partial \theta^2}\right|_{\theta=\pi} = -ml^2\omega^2 - mgl < 0$$

This point is unstable for all values of ω.

c) At $\omega_c > \sqrt{g/l}$, $\theta = \theta_3$, (S.1.32.2) becomes

$$\left.\frac{\partial^2 U_{\text{eff}}}{\partial \theta^2}\right|_{\theta=\theta_3} = -2m\omega^2 l^2 \frac{g^2}{\omega^4 l^2} + m\omega^2 l^2 + mgl\frac{g}{\omega^2 l}$$

$$= -\frac{mg^2}{\omega^2} + m\omega^2 l^2 = \begin{cases} > 0, & \omega^2 > g/l \\ < 0, & \omega^2 < g/l \end{cases}$$

So, here at $\omega_c > \sqrt{g/l}$, the equilibrium point is stable.

d) Consider the initial differential equation (S.1.32.1) and substitute for θ, $\theta \to \theta_3 + \delta$, where $\theta_3 = \cos^{-1}\left(g/\omega^2 l\right)$:

$$ml^2\left(\ddot{\theta}_3 + \ddot{\delta}\right) - m\omega^2 l^2 \sin\left(\theta_3 + \delta\right) \cdot \cos\left(\theta_3 + \delta\right) + mgl \sin\left(\theta_3 + \delta\right) = 0$$

For small oscillations, we will use the approximations $\sin\delta = \delta$, $\cos\delta = 1 - (1/2)\delta^2$ and leave only terms linear in δ :

$$ml^2\ddot{\delta} + \left[m\omega^2 l^2\left(1 - 2\cos^2\theta_3\right) + mgl\cos\theta_3\right] \cdot \delta = 0$$

After substituting $\cos\theta_3 = g/l\omega^2$, we will have for the frequency Ω of small oscillations about this point

$$ml^2\ddot{\delta} + \left(m\omega^2 l^2 - \frac{mg^2}{\omega^2}\right)\delta = 0$$

$$\Omega = \sqrt{\omega^2 - \left(g^2/\omega^2 l^2\right)}$$

1.33   Flyball Governor (Boston, Princeton, MIT)

a) Find the Lagrangian of the system. The kinetic energy

$$T = \frac{M}{2}\dot{y}^2 + \frac{1}{2}\left(2ml^2\right)\dot{\theta}^2 + \frac{1}{2}\left(2ml^2\sin^2\theta\right)\omega^2 \qquad (S.1.33.1)$$

where y is the distance of the sleeves from each other and θ is the angle of the hinged rods to the fixed vertical rod (see Figure S.1.33). The potential energy

$$V = -Mgy - 2mg\frac{y}{2} \qquad (S.1.33.2)$$

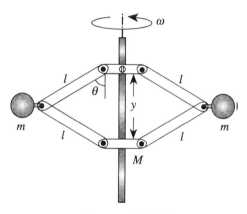

Figure S.1.33

Using the relation $y = 2l\cos\theta$, we obtain for $\theta \neq 0$

$$L = \left(\frac{M}{2} + \frac{ml^2}{4l^2 - y^2}\right)\dot{y}^2 + ml^2\left(1 - \frac{y^2}{4l^2}\right)\omega^2 + (m + M)gy \qquad (S.1.33.3)$$

The equation of motion becomes

$$\frac{d}{dt}\left[2\dot{y}\left(\frac{M}{2} + \frac{ml^2}{4l^2 - y^2}\right)\right] - \frac{2ml^2y\dot{y}^2}{(4l^2 - y^2)^2} + \frac{my\omega^2}{2} - (m+M)g = 0$$

$$(S.1.33.4)$$

b) From (S.1.33.3), we may introduce the effective potential energy

$$U_{\text{eff}} = \frac{m\omega^2}{4}y^2 - (m + M)gy \qquad (S.1.33.5)$$

Its minimum gives the equilibrium position of the sleeve y_0

$$\left.\frac{\partial U_{\text{eff}}}{\partial y}\right|_{y=y_0} = \frac{m\omega^2 y_0}{2} - (m+M)g = 0 \qquad (S.1.33.6)$$

$$y_0 = \frac{2(m+M)g}{m\omega^2} \qquad (S.1.33.7)$$

The angle θ_0 corresponding to (S.1.33.7) is defined by

$$\cos\theta_0 = \frac{y_0}{2l} = \frac{(m+M)g}{m\omega^2 l} \qquad (S.1.33.8)$$

The condition for stability of the equilibrium of (S.1.33.8) is equivalent to

$$\cos\theta_0 = \frac{(m+M)g}{m\omega^2 l} < 1 \qquad (S.1.33.9)$$

which can clearly be seen if we write everything in terms of θ and not y (see Problem 1.32). On the other hand, if

$$\frac{(m+M)g}{m\omega^2 l} > 1$$

(S.1.33.7) and (S.1.33.8) are no longer valid, and $y_0 = 2l$. This corresponds to the stable equilibrium at $\theta = 0$ (again compare to Problem 1.32). So the height z_0 of the lower sleeve above its lowest point is

$$z_0 = \begin{cases} 2l - y_0 = 2l - \dfrac{2(m+M)}{m\omega^2}g, & \dfrac{(m+M)g}{m\omega^2 l} < 1 \\[4mm] 0, & \dfrac{(m+M)g}{m\omega^2 l} > 1 \end{cases} \qquad (S.1.33.10)$$

c) Taking the time derivative in (S.1.33.4), we obtain

$$\left(M + \frac{2ml^2}{4l^2 - y^2}\right)\ddot{y} + \frac{2ml^2 y}{(4l^2 - y^2)^2}\dot{y}^2 + \frac{my}{2}\omega^2 - (m+M)g = 0 \qquad (S.1.33.11)$$

For small oscillations around the equilibrium point $y = y_0 + \eta$, the quadratic terms of η may be neglected, and we rewrite (S.1.33.11), where y_0 is defined in (S.1.33.10) under the conditions in (S.1.33.9)

$$y_0 = \frac{2g(m+M)}{m\omega^2} \qquad (S.1.33.12)$$

So
$$\ddot{\eta} + \Omega^2 \eta = 0 \qquad (S.1.33.13)$$
where Ω is the angular frequency of a simple harmonic oscillator given by

$$\Omega = \sqrt{\frac{m\omega^2/2}{M + 2ml^2/(4l^2 - y_0^2)}} \qquad (S.1.33.14)$$

Now, $y_0 = 2l \cos \theta_0$, and using (S.1.33.9), we eliminate ω^2 to arrive at

$$\Omega = \sqrt{\frac{(m + M)g \sin^2 \theta_0}{(m + 2M \sin^2 \theta_0)l \cos \theta_0}} \qquad (S.1.33.15)$$

1.34   Double Pendulum (Stony Brook, Princeton, MIT)

a) For the first mass m, the Lagrangian is given by

$$\mathcal{L}_1 = T_1 - V_1 = \frac{1}{2}ml^2\dot{\theta}_1^2 - mgl\,(1 - \cos\theta_1) = \frac{1}{2}ml^2\dot{\theta}_1^2 + mgl\cos\theta_1$$

ignoring the constant mgl. To find $\mathcal{L}_2$, introduce the coordinates for the second mass (see Figure S.1.34):

$$x_2 = l\sin\theta_1 + l\sin\theta_2$$

$$y_2 = l\cos\theta_1 + l\cos\theta_2$$

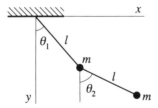

Figure S.1.34

Now, $\mathcal{L}_2 = T_2 - V_2$, where

$$T_2 = \frac{m}{2}\left(\dot{x}_2^2 + \dot{y}_2^2\right) = \frac{m}{2}l^2\left[\dot{\theta}_1^2 + \dot{\theta}_2^2 + 2\dot{\theta}_1\dot{\theta}_2\cos\left(\theta_2 - \theta_1\right)\right]$$

$$V_2 = -mgy_2 = -mgl\,(\cos\theta_1 + \cos\theta_2)$$

So

$$\mathcal{L} = \mathcal{L}_1 + \mathcal{L}_2$$

$$= ml^2\dot{\theta}_1^2 + \frac{m}{2}l^2\dot{\theta}_2^2 + ml^2\dot{\theta}_1\dot{\theta}_2\cos(\theta_2 - \theta_1) + 2mgl\cos\theta_1 + mgl\cos\theta_2$$

For $\theta_1, \theta_2 \ll 1$, we can take $\cos\theta = 1 - \theta^2/2$. Denoting the frequency of a single pendulum by $\omega_0 = \sqrt{g/l}$, and eliminating superfluous constant terms, we obtain the Lagrangian in the form

$$\mathcal{L} \approx ml^2\left(\dot{\theta}_1^2 + \frac{1}{2}\dot{\theta}_2^2 + \dot{\theta}_1\dot{\theta}_2 - \omega_0^2\theta_1^2 - \frac{1}{2}\omega_0^2\theta_2^2\right) \qquad (S.1.34.1)$$

b) Using (S.1.34.1) we can write the equations of motion

$$2\ddot{\theta}_1 + \ddot{\theta}_2 + 2\omega_0^2\theta_1 = 0$$

$$\ddot{\theta}_1 + \ddot{\theta}_2 + \omega_0^2\theta_2 = 0$$

$$(S.1.34.2)$$

c) We are looking for solutions of (S.1.34.2) of the form

$$\theta_1 = Ae^{i\omega t}$$

$$\theta_2 = Be^{i\omega t}$$

$$(S.1.34.3)$$

After substituting (S.1.34.3) into (S.1.34.2), we get a pair of linear equations in A and B

$$-2A\omega^2 - B\omega^2 + 2A\omega_0^2 = 0$$

$$-A\omega^2 - B\omega^2 + B\omega_0^2 = 0$$

$$(S.1.34.4)$$

For nontrivial solutions of (S.1.34.4) to exist, we should have

$$\det \begin{vmatrix} 2\omega_0^2 - 2\omega^2 & -\omega^2 \\ -\omega^2 & \omega_0^2 - \omega^2 \end{vmatrix} = 0 \qquad (S.1.34.5)$$

The eigenfrequencies are defined from

$$2\left(\omega_0^2 - \omega^2\right)^2 = \omega^4 \qquad \sqrt{2}\left(\omega_0^2 - \omega^2\right) = \pm\omega^2 \qquad (S.1.34.6)$$

Finally,

$$\omega_1^2 = \frac{\sqrt{2}}{\sqrt{2}+1}\,\omega_0^2 \qquad \omega_1 = \left(2-\sqrt{2}\right)^{1/2}\omega_0$$

$$\omega_2^2 = \frac{\sqrt{2}}{\sqrt{2}-1}\,\omega_0^2 \qquad \omega_2 = \left(2+\sqrt{2}\right)^{1/2}\omega_0$$

(S.1.34.7)

1.35   Triple Pendulum (Princeton)

a) Write the Lagrangian of the system using coordinates $\varphi_1, \varphi_2, \varphi_3$ (see Figure S.1.35a).

$$x_1 = a\sin\varphi_1 \quad x_2 = a(\sin\varphi_1 + \sin\varphi_2) \quad x_3 = a(\sin\varphi_1 + \sin\varphi_2 + \sin\varphi_3)$$

$$y_1 = a\cos\varphi_1 \quad y_2 = a(\cos\varphi_1 + \cos\varphi_2) \quad y_3 = a(\cos\varphi_1 + \cos\varphi_2 + \cos\varphi_3)$$

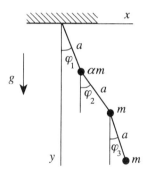

Figure S.1.35a

Then in the small angle approximation,

$$T = \frac{\alpha m}{2}\left(\dot{x}_1^2 + \dot{y}_1^2\right) + \frac{m}{2}\left(\dot{x}_2^2 + \dot{y}_2^2\right) + \frac{m}{2}\left(\dot{x}_3^2 + \dot{y}_3^2\right)$$

$$\approx \frac{\alpha m a^2}{2}\,\dot{\varphi}_1^2 + \frac{ma^2}{2}\left(\dot{\varphi}_1^2 + \dot{\varphi}_2^2 + 2\dot{\varphi}_1\dot{\varphi}_2\right)$$

$$+\frac{ma^2}{2}\left(\dot{\varphi}_1^2 + \dot{\varphi}_2^2 + \dot{\varphi}_3^2 + 2\dot{\varphi}_1\dot{\varphi}_2 + 2\dot{\varphi}_2\dot{\varphi}_3 + 2\dot{\varphi}_1\dot{\varphi}_3\right)$$

$$V = -\alpha mgy_1 - mgy_2 - mgy_3$$

$$\approx \frac{1}{2}\, mga\alpha\varphi_1^2 + \frac{1}{2}\, mga(\varphi_1^2 + \varphi_2^2) + \frac{1}{2}\, mga(\varphi_1^2 + \varphi_2^2 + \varphi_3^2)$$

Here we used

$$\sin \varphi_i \approx \varphi_i$$

$$\cos \varphi_i \cos \varphi_j + \sin \varphi_i \sin \varphi_j = \cos(\varphi_i - \varphi_j) \approx 1$$

So the Lagrangian is

$$\mathcal{L} = T - V = \frac{ma^2}{2} \left\{ (\alpha + 2)\dot{\varphi}_1^2 + 2\dot{\varphi}_2^2 + \dot{\varphi}_3^2 + 4\dot{\varphi}_1\dot{\varphi}_2 + 2\dot{\varphi}_1\dot{\varphi}_3 + 2\dot{\varphi}_2\dot{\varphi}_3 \right.$$

$$\left. -\omega_0^2 \left[(\alpha + 2)\varphi_1^2 + 2\varphi_2^2 + \varphi_3^2 \right] \right\}$$

where we let $g/a = \omega_0^2$. Therefore the equations of motion will be

$$(\alpha + 2)\left[\ddot{\varphi}_1 + \omega_0^2\varphi_1\right] + 2\ddot{\varphi}_2 + \ddot{\varphi}_3 = 0$$

$$2\ddot{\varphi}_1 + 2\ddot{\varphi}_2 + 2\omega_0^2\varphi_2 + \ddot{\varphi}_3 = 0$$

$$\ddot{\varphi}_1 + \ddot{\varphi}_2 + \ddot{\varphi}_3 + \omega_0^2\varphi_3 = 0$$

Looking for the solution in the form $\varphi_i = A_i e^{i\omega t}$ and letting $\omega^2/\omega_0^2 = \lambda$, we have as a condition for the existence of a nontrivial solution

$$\det \begin{vmatrix} (\alpha + 2)(\lambda - 1) & 2\lambda & \lambda \\ 2\lambda & 2(\lambda - 1) & \lambda \\ \lambda & \lambda & \lambda - 1 \end{vmatrix} = 0$$

We want a mode where $\omega^2 = 2g/a$. So $\lambda = 2$, and the determinant becomes

$$\det \begin{vmatrix} \alpha + 2 & 4 & 2 \\ 4 & 2 & 2 \\ 2 & 2 & 1 \end{vmatrix} = 0$$

Obviously $\alpha = 2$ is the only solution of this equation (the first and third rows are then proportional).

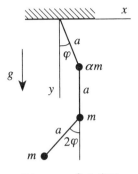

Figure S.1.35b

b) The mode corresponding to this frequency can be found from the equation

$$\begin{pmatrix} 4 & 4 & 2 \\ 4 & 2 & 2 \\ 2 & 2 & 1 \end{pmatrix} \begin{pmatrix} A_1 \\ A_2 \\ A_3 \end{pmatrix} = 0$$

which has a solution $A_3 = -2A_1$, $A_2 = 0$. So the mode corresponding to the frequency $\omega = \sqrt{2g/a}$ is shown in Figure S.1.35b

$$\begin{pmatrix} \varphi_1 \\ \varphi_2 \\ \varphi_3 \end{pmatrix} \sim \begin{pmatrix} \varphi \\ 0 \\ -2\varphi \end{pmatrix}$$

1.36 Three Masses and Three Springs on Hoop (Columbia, Stony Brook, MIT)

Introducing x_i, the displacement from equilibrium for respective masses 1,2,3 (see Figure S.1.36), we can write a Lagrangian in the form

$$\mathcal{L} = T - V = \frac{m}{2} \left(\dot{x}_1^2 + \dot{x}_2^2 + \dot{x}_3^2 \right) - \frac{k}{2} \left[(x_2 - x_1)^2 + (x_3 - x_2)^2 + (x_1 - x_3)^2 \right]$$

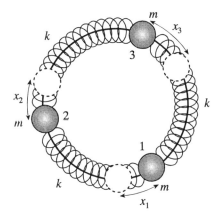

Figure **S.1.36**

The resulting equations of motion are in the form

$$\ddot{x}_1 + \frac{2k}{m}x_1 - \frac{k}{m}(x_2 + x_3) = 0$$

$$\ddot{x}_2 + \frac{2k}{m}x_2 - \frac{k}{m}(x_1 + x_3) = 0 \qquad \text{(S.1.36.1)}$$

$$\ddot{x}_3 + \frac{2k}{m}x_3 - \frac{k}{m}(x_1 + x_2) = 0$$

Again, looking for solutions of the form $x_i = A_i e^{i\omega t}$, we obtain an equation for the determinant:

$$\det \begin{vmatrix} -\lambda + 2 & -1 & -1 \\ -1 & -\lambda + 2 & -1 \\ -1 & -1 & -\lambda + 2 \end{vmatrix} = 0$$

where $\lambda = \omega^2/\omega_0^2$ and $\omega_0^2 = k/m$. The first root is $\lambda_1 = \omega = 0$ which corresponds to the movement of all three masses with the same velocity. The two other roots are $\lambda_{2,3} = 3$, $\omega = \sqrt{3}\,\omega_0$, which are double degenerate, corresponding to the mode $A_1 = 0$, $A_2 = -A_3$, or $A_2 = 0$, $A_1 = -A_3$, where one mass is at rest and the two others move in opposite directions. The result can be obtained even without solving (S.1.36.1), if one can guess that this is the mode. Then

$$T = \frac{m}{2}\dot{x}^2 + \frac{m}{2}\dot{x}^2 = m\dot{x}^2$$

$$V = 2 \cdot \frac{k}{2}x^2 + \frac{k}{2}(2x)^2 = 3kx^2$$

So, again, $\omega = \sqrt{3k/m} = \sqrt{3}\,\omega_0$.

1.37 Nonlinear Oscillator (Princeton)

a) The Lagrangian for the potential $V(x)$ is

$$\mathcal{L} = T - V(x) = \frac{1}{2}m\dot{x}^2 - \frac{1}{2}kx^2 + \frac{1}{3}m\lambda x^3$$

Therefore, the equation of motion for the nonlinear harmonic oscillator is

$$\ddot{x} + \omega_0^2 x = \lambda x^2 \tag{S.1.37.1}$$

where $\omega_0 = \sqrt{k/m}$ is the principal frequency of a harmonic oscillator. We will look for a solution of the form

$$x = x^{(0)} + \lambda x^{(1)} + \lambda^2 x^{(2)} + \cdots \tag{S.1.37.2}$$

where $x^{(0)}$ is a solution of a harmonic oscillator equation

$$\ddot{x}^{(0)} + \omega_0^2 x^{(0)} = 0 \tag{S.1.37.3}$$

Since we are looking only for the first order corrections, we do not have to consider a frequency shift in the principal frequency ω_0. The solution of equation (S.1.37.3) with initial condition $x^{(0)} = 0$ is $x^{(0)} = A\sin\omega_0 t$. Substituting this into (S.1.37.1) and using (S.1.37.2), we obtain an equation for $x^{(1)}$ (leaving only the terms which are first order in λ).

$$\lambda\ddot{x}^{(1)} + \lambda\omega_0^2 x^{(1)} = \lambda\left(x^{(0)} + \lambda x^{(1)}\right)^2 = \lambda x^{(0)^2} \tag{S.1.37.4}$$

or

$$\ddot{x}^{(1)} + \omega_0^2 x^{(1)} = x^{(0)^2} = A^2 \sin^2 \omega_0 t = \frac{A^2}{2}(1 - \cos 2\omega_0 t) \tag{S.1.37.5}$$

The solution for $x^{(1)}$ is a sum of the solutions of the linear homogeneous and the linear inhomogeneous equations:

$$x^{(1)} = B\sin\omega_0 t + C\cos\omega_0 t + \tilde{x} \tag{S.1.37.6}$$

where $\tilde{x}$ is the inhomogeneous solution of the form

$$\tilde{x} = \frac{A^2}{2\omega_0^2} + D\cos 2\omega_0 t \qquad (S.1.37.7)$$

Substituting (S.1.37.6) and (S.1.37.7) into (S.1.37.5), we obtain $D = A^2/6\omega_0^2$. So

$$x^{(1)} = B\sin\omega_0 t + C\cos\omega_0 t + \frac{A^2}{2\omega_0^2} + \frac{A^2}{6\omega_0^2}\cos 2\omega_0 t \qquad (S.1.37.8)$$

Using the initial condition $x^{(1)}(0) = 0$ we obtain $C = -2A^2/3\omega_0^2$. The solution of the equation of motion (S.1.37.1) will be

$$x = A'\sin\omega_0 t - \frac{2}{3}\frac{A^2\lambda}{\omega_0^2}\cos\omega_0 t + \frac{1}{2}\frac{A^2\lambda}{\omega_0^2} + \frac{1}{6}\frac{A^2\lambda}{\omega_0^2}\cos 2\omega_0 t \qquad (S.1.37.9)$$

where A' is defined from initial conditions.

b) The average of $\langle x\rangle$ over a period $T = 2\pi/\omega_0$ is certainly nonzero for a given amplitude of oscillation A. Inspection of (S.1.37.9) reveals that

$$\langle x\rangle = \frac{1}{2}\frac{A^2\lambda}{\omega_0^2} \qquad (S.1.37.10)$$

To take into account the energy distribution of the amplitude, we have to calculate the thermodynamical average of $\langle x\rangle$ as a function of temperature

$$\overline{\langle x\rangle} = \frac{\int\limits_0^\infty \langle x\rangle e^{-\varepsilon/\tau}\,d\varepsilon}{\int\limits_0^\infty e^{-\varepsilon/\tau}\,d\varepsilon} \qquad (S.1.37.11)$$

where τ is the temperature in energy units $\tau = k_B T$ and k_B is Boltzmann's constant.

$$\overline{\langle x\rangle} = \frac{\lambda}{2\omega_0^2\tau}\int\limits_0^\infty A^2(\varepsilon)e^{-\varepsilon/\tau}\,d\varepsilon \qquad (S.1.37.12)$$

The amplitude of the oscillator as a function of energy is given by

$$A^2 = \frac{2\varepsilon}{k} \qquad (S.1.37.13)$$

Substituting (S.1.37.13) into (S.1.37.12) gives

$$\overline{\langle x\rangle} = \frac{\lambda}{k\omega_0^2\tau}\int\limits_0^\infty \varepsilon e^{-\varepsilon/\tau}\,d\varepsilon = \frac{\lambda}{k\omega_0^2}\tau \qquad (S.1.37.14)$$

This result can explain the nonzero thermal expansion coefficient of solids. As τ increases, the equilibrium point shifts (see also the discussion in Kittel, *Introduction to Solid State Physics*, p.142, where (S.1.37.14) is obtained by a different method).

1.38 Swing (MIT, Moscow Phys-Tech)

Consider half a period of swinging motion between points $0 \to 1 \to 2 \to 3 \to 4$ (see Figure S.1.38, where the dotted line indicates the position of the center of mass). From 0 to 1, energy is conserved $E_0 = E_1$.

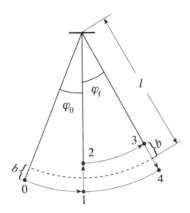

Figure S.1.38

$$mg\,(l + b)\,(1 - \cos\varphi_0) = \frac{mv_1^2}{2}$$

where φ_0 is the initial angle. For small angles, $1 - \cos\varphi_0 = \varphi_0^2/2$ and

$$mg\,(l + b)\,\frac{\varphi_0^2}{2} = \frac{mv_1^2}{2} \qquad\qquad (S.1.38.1)$$

From 1 to 2, angular momentum is conserved:

$$m\,(l + b)\,v_1 = m\,(l - b)\,v_2$$

$$(S.1.38.2)$$

$$v_2 = \left(\frac{l + b}{l - b}\right) v_1$$

From 2 to 3, again, energy is conserved so we can find the final angle φ_f

$$mg\left(l-b\right)\frac{\varphi_f^2}{2} = \frac{mv_2^2}{2} = \frac{mv_1^2}{2}\left(\frac{l+b}{l-b}\right)^2 \qquad (S.1.38.3)$$

From (S.1.38.1) and (S.1.38.3) we can express $mv_1^2/2$ and obtain the change in the amplitude:

$$mg\left(l+b\right)\varphi_0^2 = mg\left(l-b\right)\left(\frac{l-b}{l+b}\right)^2\varphi_f^2 \qquad (S.1.38.4)$$

or using $b/l \ll 1$

$$\varphi_f \approx \left(1 + 3\frac{b}{l}\right)\varphi_0 \qquad (S.1.38.5)$$

and

$$\Delta\varphi = 3\frac{b}{l}\varphi_0 \qquad (S.1.38.6)$$

The work done by the child is equal to the energy change per period :

$$\Delta E = 2 \cdot \frac{mgl}{2}\left(\varphi_f^2 - \varphi_0^2\right) \approx 2mgl\varphi_0\Delta\varphi$$
$$= 6mgb\varphi_0^2 = 12E_0\frac{b}{l} \qquad (S.1.38.7)$$

We want to write (S.1.38.7) in the form $dE/dt = \alpha E$,

$$\frac{dE}{dt} \approx \frac{\Delta E}{2\pi/\omega_0} \approx 6\frac{\omega_0}{\pi}\frac{b}{l}E_0 \qquad (S.1.38.8)$$

$$\alpha = \frac{6\omega_0}{\pi}\frac{b}{l} \qquad (S.1.38.9)$$

where $\omega_0 = \sqrt{g/l}$.

1.39   Rotating Door (Boston)

We will use the frame rotating with the door (body frame, axes $\hat{\mathbf{1}}, \hat{\mathbf{2}}, \hat{\mathbf{3}}$, see Figure S.1.39), so that we can use the Euler equations

$$I_1\frac{d\omega_1}{dt} + (I_3 - I_2)\,\omega_2\omega_3 = N_1 \qquad (S.1.39.1)$$

$$I_2\frac{d\omega_2}{dt} + (I_1 - I_3)\,\omega_1\omega_3 = N_2 \qquad (S.1.39.2)$$

$$I_3\frac{d\omega_3}{dt} + (I_2 - I_1)\,\omega_1\omega_2 = N_3 \qquad (S.1.39.3)$$

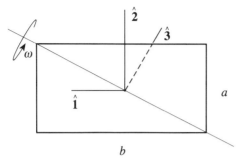

Figure S.1.39

where $\omega_1, \omega_2, \omega_3$ are the components of angular velocity in this frame and I_1, I_2, I_3 are the principal moments of inertia. In this case I_1 and I_2 correspond to the moments of inertia of rods of length a and b, respectively, which can be calculated easily (see the solution to Problem 1.30):

$$I_1 = \frac{1}{12}ma^2 \qquad I_2 = \frac{1}{12}mb^2$$

Since the problem is two-dimensional (we can disregard the thickness of the door),

$$I_3 = I_1 + I_2 = \frac{1}{12}m\left(a^2 + b^2\right)$$

In our frame $\omega_3 = 0$, and ω_1 and ω_2 are constant.

$$\omega_1 = \frac{\omega b}{\sqrt{a^2 + b^2}} \qquad\qquad\qquad (S.1.39.4)$$

$$\omega_2 = \frac{\omega a}{\sqrt{a^2 + b^2}} \qquad\qquad\qquad (S.1.39.5)$$

Substituting (S.1.39.4) and (S.1.39.5) into (S.1.39.1)–(S.1.39.3) we obtain

$$N_1 = N_2 = 0$$

$$|\mathbf{N}| = N_3 = \frac{m\left(b^2 - a^2\right)ab\omega^2}{12\left(a^2 + b^2\right)}$$

1.40   Bug on Globe (Boston)

The angular velocity of the globe is always in the same direction (along the fixed axis, see Figure S.1.40). Since the angular momentum $\mathbf{l}$ is constant

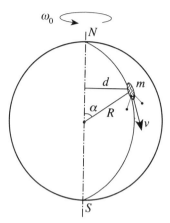

Figure **S.1.40**

and $|\mathbf{l}| = I\omega$ we may write

$$\frac{d}{dt}(I\omega) = 0$$

$$I\omega = \text{const} = I_0\omega_0 \tag{S.1.40.1}$$

Initially I_0 is just the moment of inertia of the sphere (the bug is at the pole), so $I_0 = (2/5)MR^2$. Substituting this into (S.1.40.1) we obtain

$$I(t)\frac{d\theta}{dt} = \frac{2}{5}MR^2\omega_0 \tag{S.1.40.2}$$

where $I(t) = I_0 + md^2 = I_0 + mR^2\sin^2\alpha = I_0 + mR^2\sin^2(vt/R)$, $\alpha \equiv vt/R$, so

$$\Delta\theta = \int_0^T \frac{(2/5)MR^2\omega_0}{I(t)}\,dt = \int_0^T \frac{(2/5)MR^2\omega_0\,dt}{(2/5)MR^2 + mR^2\sin^2(vt/R)}$$

$$= \frac{R\omega_0}{v}\int_0^T \frac{d(2vt/R)}{2 + (5m/M)\sin^2(vt/R)} = \frac{R\omega_0}{v}\int_0^{2\pi} \frac{dx}{2 + (5m/M)\sin^2(x/2)}$$

$$= \frac{R\omega_0}{v}\int_0^{2\pi} \frac{dx}{(2 + 5m/2M) - (5m/2M)\cos x} = \frac{\pi\omega_0 R}{v}\sqrt{2M/(2M + 5m)}$$

We used here $\sin^2(x/2) = (1/2)(1 - \cos x)$ and the integral given in the problem:

$$\int_0^{2\pi} \frac{dx}{a + b\cos x} = \frac{2\pi}{\sqrt{a^2 - b^2}}, \qquad (a^2 > b^2)$$

If the bug had mass $m = 0$, the angle $\Delta\theta$ would be

$$\Delta\theta = \frac{\pi\omega_0 R}{v} = \omega_0\frac{\pi R}{v} = \omega_0 T$$

which corresponds to the free rotation of the globe with angular velocity ω_0.

1.41 Rolling Coin (Princeton, Stony Brook)

We can use the standard method of Euler equations to solve this problem. However, since the coin has a symmetry axis, it is easier to use a frame of reference rotating with angular velocity Ω, corresponding to the rotation of the center of mass of the coin. Rolling without slipping implies that the velocity of the point of contact with the table should be zero, and therefore

$$\omega a = \Omega(b + a\sin\theta) \tag{S.1.41.1}$$

where ω is the angular velocity of rotation of the coin around its axis (see Figure S.1.41). We have in this frame

$$\mathbf{n} = -\mathbf{w} = mg\hat{\mathbf{z}}$$

$$\mathbf{f}_{\text{fr}} = -m\Omega^2 b\hat{\mathbf{y}} \tag{S.1.41.2}$$

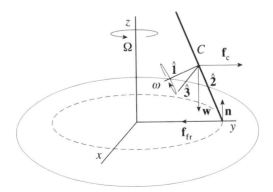

Figure **S.1.41**

where **n** is the normal reaction, and **w** is the weight. From (S.1.41.2), we can find the torque **N** relative to the center of mass of the coin:

$$\mathbf{N} = \sum_i \mathbf{r}_i \times \mathbf{F}_i = \left[mga \sin\theta - m\Omega^2 ab \cos\theta \right] \hat{\mathbf{x}} \qquad \text{(S.1.41.3)}$$

In a noninertial frame rotating with an angular velocity Ω, we can write (see for instance Goldstein, *Classical Mechanics*, §4.9)

$$\mathbf{N} = \left(\frac{d\mathbf{l}}{dt} \right)_{\text{space}} = \left(\frac{\delta\mathbf{l}}{\delta t} \right)_{\text{rot}} + \Omega \times \mathbf{l} \qquad \text{(S.1.41.4)}$$

In the rotating frame

$$\left(\frac{\delta\mathbf{l}}{\delta t} \right)_{\text{rot}} = 0$$

Choose the moment of time when one axis is horizontal and in the same direction as $\hat{\mathbf{x}}$ ($\hat{\mathbf{3}}$ in Figure S.1.41). Since the axis of this frame coincides with the principal axis, the tensor of inertia is diagonal

$$\tilde{I} = \begin{pmatrix} \dfrac{1}{2}ma^2 & 0 & 0 \\[2mm] 0 & \dfrac{1}{4}ma^2 & 0 \\[2mm] 0 & 0 & \dfrac{1}{4}ma^2 \end{pmatrix}$$

To calculate this tensor, we used $I_1 = (1/2)ma^2$ for the moment of inertia of the disk about its symmetry axis and also the fact that for a body of negligible thickness $I_1 = I_2 + I_3$. On the other hand $\mathbf{l} = \tilde{I}\Omega_{\text{tot}}$, where

$$\Omega_{\text{tot}} = (\omega - \Omega\sin\theta, -\Omega\cos\theta, 0)$$

Taking **N** from (S.1.41.4) we have

$$N\hat{\mathbf{x}} = N\hat{\mathbf{3}} = \begin{vmatrix} \hat{\mathbf{1}} & \hat{\mathbf{2}} & \hat{\mathbf{3}} \\[2mm] -\Omega\sin\theta & -\Omega\cos\theta & 0 \\[2mm] \dfrac{1}{2}ma^2(\omega - \Omega\sin\theta) & -\dfrac{1}{4}ma^2\Omega\cos\theta & 0 \end{vmatrix}$$

$$= ma^2\Omega^2 \left[\frac{1}{4}\sin\theta\cos\theta + \frac{1}{2}\left(\frac{\omega}{\Omega} - \sin\theta \right)\cos\theta \right] \hat{\mathbf{3}} \qquad \text{(S.1.41.5)}$$

From (S.1.41.1)

$$\omega = \Omega \left(\frac{b}{a} + \sin \theta \right) \qquad (S.1.41.6)$$

Comparing (S.1.41.5) with (S.1.41.3) and using (S.1.41.6) we obtain

$$ga \sin \theta - \Omega^2 ab \cos \theta = a^2 \Omega^2 \left[\frac{1}{4} \sin \theta \cos \theta + \frac{1}{2} \frac{b}{a} \cos \theta \right] \qquad (S.1.41.7)$$

$$g \sin \theta = \Omega^2 \left[\frac{a}{4} \sin \theta \cos \theta + \frac{3}{2} b \cos \theta \right] \qquad (S.1.41.8)$$

$$\Omega^2 = \frac{4g \tan \theta}{6b + a \sin \theta} \qquad (S.1.41.9)$$

1.42   Unstable Top (Stony Brook)

a) There are two integrals of motion in the generalized momenta p_φ, p_ψ :

$$p_\varphi = \frac{\partial L}{\partial \dot{\varphi}} = I_3 \left(\dot{\psi} + \dot{\varphi} \cos \theta \right) \cos \theta + I_1 \dot{\varphi} \sin^2 \theta \qquad (S.1.42.1)$$

$$p_\psi = \frac{\partial L}{\partial \dot{\psi}} = I_3 \left(\dot{\psi} + \dot{\varphi} \cos \theta \right) = I_3 \omega_3 \qquad (S.1.42.2)$$

where we used the fact that $\omega_3 = \dot{\psi} + \dot{\varphi} \cos \theta$ is the angular velocity of the top around its axis. Applying the initial conditions ($\theta = 0$, $\dot{\theta} = 0$) to (S.1.42.1) and (S.1.42.2), we obtain

$$p_\varphi = p_\psi = I_3 \omega_3$$

Another integral of motion is, of course, the energy; again using the initial conditions, we have

$$E = T + V = \frac{1}{2} I_1 \left(\dot{\theta}^2 + \dot{\varphi}^2 \sin^2 \theta \right) + \frac{1}{2} I_3 \omega_3^2 + M g l \cos \theta$$

$$= E_0 = \frac{1}{2} I_3 \omega_3^2 + M g l \qquad (S.1.42.3)$$

b) From (S.1.42.3) and using the condition that the head will descend to a maximum angle θ where $\dot{\theta} = 0$, we have

$$\dot{\varphi}^2 = \frac{2 M g l \left(1 - \cos \theta \right)}{I_1 \sin^2 \theta} \qquad (S.1.42.4)$$

On the other hand, from (S.1.42.1),

$$\dot{\varphi} = \frac{I_3\omega_3\left(1 - \cos\theta\right)}{I_1 \sin^2\theta} \qquad (S.1.42.5)$$

By equating $\dot{\varphi}$ in (S.1.42.4) and (S.1.42.5) and using the half angle formulas

$$1 - \cos\theta = 2\sin^2\frac{\theta}{2} \qquad \sin\theta = 2\sin\frac{\theta}{2}\cos\frac{\theta}{2}$$

we get

$$\cos\frac{\theta}{2} = \frac{I_3\omega_3}{2\sqrt{I_1 M g l}}$$

c) Again using (S.1.42.3) and (S.1.42.5), we have

$$\dot{\theta}^2 = \frac{2}{I_1}Mgl\left(1 - \cos\theta\right) - \dot{\varphi}^2 \sin^2\theta = \frac{4Mgl}{I_1}\sin^2\frac{\theta}{2} - \frac{I_3^2\omega_3^2}{I_1^2}\tan^2\frac{\theta}{2}$$

1.43 Pendulum Clock in Noninertial Frame (Maryland)

Calculate the Lagrangian of the mass m and derive the equation of motion for $\theta\left(t\right)$ (see Figure S.1.43). Start with the equations for the x and y positions of the mass

$$x = R\cos\omega t + l\cos\left(\theta + \omega t\right)$$

$$y = R\sin\omega t + l\sin\left(\theta + \omega t\right)$$

and compose

$$\mathcal{L} = \frac{1}{2}m\left(\dot{x}^2 + \dot{y}^2\right) = \frac{1}{2}m\left[\omega^2 R^2 + l^2\left(\dot{\theta} + \omega\right)^2 + 2Rl\omega\left(\dot{\theta} + \omega\right)\cos\theta\right].$$

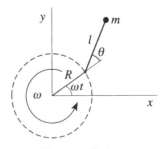

Figure **S.1.43**

Applying Lagrange's equations gives

$$\ddot{\theta} + \frac{R}{l}\omega^2 \sin\theta = 0$$

which, for $g = \omega^2 R$, corresponds, as required, to the equation of motion for a pendulum $\ddot{\theta} + (g/l)\sin\theta = 0$ in a uniform gravitational field.

1.44 Beer Can (Princeton, Moscow Phys-Tech)

a) It is possible to solve this problem in the rotating frame of the station. However, then you have to take into account the Coriolis force. Instead, consider the movement of the can in the Earth's frame first and assume that the trajectory of the can is only slightly perturbed compared to the trajectory of the station (see, for instance, Arnold, *Mathematical Methods of Classical Mechanics*, §2.8). Write Newton's equations for the orbit

$$\ddot{\mathbf{r}} = -\Omega^2 \frac{\mathbf{r}}{r^3} \tag{S.1.44.1}$$

where Ω is the frequency of revolution. Using polar coordinates (r, φ), we can write for the trajectory of the can

$$r = r_0 + r_1$$
$$\varphi = \varphi_0 + \varphi_1 \tag{S.1.44.2}$$

where (r_0, φ_0) correspond to the trajectory of the station and (r_1, φ_1) are small corrections to the trajectory. Writing

$$\mathbf{r} = r\mathbf{e}_r$$

$$\dot{\mathbf{e}}_r = \dot{\varphi}\mathbf{e}_\varphi$$

$$\dot{\mathbf{e}}_\varphi = -\dot{\varphi}\mathbf{e}_r$$

we have

$$\ddot{\mathbf{r}} = \ddot{r}\mathbf{e}_r + \dot{r}\dot{\varphi}\mathbf{e}_\varphi + \dot{r}\dot{\varphi}\mathbf{e}_\varphi + r\ddot{\varphi}\mathbf{e}_\varphi - r\dot{\varphi}^2\mathbf{e}_r \tag{S.1.44.3}$$

Substituting (S.1.44.3) into (S.1.44.1) yields

$$\ddot{r} - r\dot{\varphi}^2 + \frac{1}{r^2} = 0 \tag{S.1.44.4}$$

$$2\dot{r}\dot{\varphi} + r\ddot{\varphi} = 0 \tag{S.1.44.5}$$

where we have chosen units of time and length so that

$$r_0 = 1 \qquad \dot{\varphi}_0 = \Omega = 1 \qquad\qquad \text{(S.1.44.6)}$$

Substituting the variables for the perturbed orbit into (S.1.44.4) and (S.1.44.5) from (S.1.44.2) gives

$$\ddot{r}_0 + \ddot{r}_1 - (r_0 + r_1)(\dot{\varphi}_0 + \dot{\varphi}_1)^2 + \frac{1}{(r_0 + r_1)^2} = 0 \qquad \text{(S.1.44.7)}$$

$$2(\dot{r}_0 + \dot{r}_1)(\dot{\varphi}_0 + \dot{\varphi}_1) + (r_0 + r_1)(\ddot{\varphi}_0 + \ddot{\varphi}_1) = 0 \qquad \text{(S.1.44.8)}$$

Using the units defined in (S.1.44.6) and assuming that $r_1 \ll 1$ and $\dot{\varphi}_1 \ll 1$ simplifies (S.1.44.7) and (S.1.44.8):

$$\ddot{r}_1 = 3r_1 + 2\dot{\varphi}_1$$
$$\text{(S.1.44.9)}$$
$$\ddot{\varphi}_1 = -2\dot{r}_1$$

where we expanded

$$\frac{1}{(r_0 + r_1)^2} \approx \frac{1}{1 + 2r_1} \approx 1 - 2r_1$$

Solving (S.1.44.9) by differentiating gives

$$\dddot{r}_1 + \dot{r}_1 = 0$$

$$\dot{r}_1 = A \sin t + B \cos t$$

where A and B are some constants, so

$$r_1(t) = -A \cos t + B \sin t \qquad\qquad \text{(S.1.44.10)}$$

In the Earth's frame, the orbit of the can is only slightly perturbed compared to the orbit of the station. In order for the beer can to appear to be rotating around the station it should have the same period T as that of the station. The period only depends on the major axis of the ellipse which was $r_0 + \epsilon$ before the can was thrown and should become r_0 after. On the other hand, the major axis only depends on the energy of the orbit.

$$a \propto \frac{1}{|E|}$$

To change the energy, we need only change the tangential velocity of the can. The initial conditions are

$$\dot{r}_1(0) = 0 \qquad \text{(S.1.44.11)}$$

$$r_1(0) = \epsilon/r_0 \qquad \text{(S.1.44.12)}$$

From (S.1.44.11) $B = 0$ and from (S.1.44.12) $A = -\epsilon/r_0$, so (S.1.44.10) becomes

$$r_1(t) = \epsilon/r_0 \cos t \qquad \text{(S.1.44.13)}$$

From (S.1.44.9)

$$\ddot{\varphi}_1 = \frac{2\epsilon}{r_0} \sin t \qquad \text{(S.1.44.14)}$$

Integrating (S.1.44.14) with the initial condition

$$\varphi_1(0) = 0$$

$$\dot{\varphi}_1(0) \equiv \omega_0$$

we obtain

$$\varphi_1 = -\frac{2\epsilon}{r_0} \sin t + (\omega_0 + \frac{2\epsilon}{r_0})t \qquad \text{(S.1.44.15)}$$

In order for the can to orbit around the station φ_1 should be periodic, i.e., the term

$$\omega_0 + \frac{2\epsilon}{r_0} = 0$$

which gives for the initial angular velocity of the can

$$\omega_0 = -\frac{2\epsilon}{r_0}$$

or for the velocity

$$V_0 = -\frac{2\epsilon}{r_0}\left(1 + \frac{\epsilon}{r_0}\right) \approx -\frac{2\epsilon}{r_0}$$

In the usual units

$$V_0 = -2\epsilon\Omega \qquad \text{(S.1.44.16)}$$

The minus sign means that the can should be thrown in the direction opposite to the direction of rotation of the station.

b) The parameters of the orbit will be defined by the equations

$$r_1(t) = \frac{\epsilon}{r_0} \cos t$$

$$\varphi_1(t) = -\frac{2\epsilon}{r_0} \sin t \qquad \text{(S.1.44.17)}$$

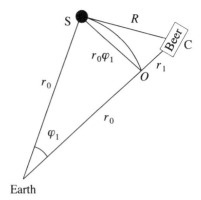

Figure **S.1.44a**

From (S.1.44.17), we can find the equation of the trajectory of the can seen by the observer in the station. The distance R from the station (S) to the can (C) (in regular units)

$$R^2 = r_1^2 + r_0^2 \varphi_1^2 \qquad (S.1.44.18)$$

where we assumed that φ_1 is small and the angle $\angle COS \approx \pi/2$ (see Figure S.1.44a). Substituting r_1 and φ_1 from (S.1.44.17) into (S.1.44.18), we obtain

$$R^2 = \epsilon^2 \cos^2 \Omega t + 4\epsilon^2 \sin^2 \Omega t = \epsilon^2 (1 + 3\sin^2 \Omega t) \qquad (S.1.44.19)$$

So in this approximation, the orbit of the can as seen from the station is an ellipse of major axis 2ϵ and minor axis ϵ with the same period T as the station. The position of the station and the can in the Earth's frame is shown in Figure S.1.44b.

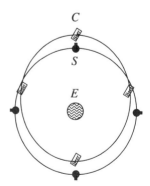

Figure **S.1.44b**

1.45   Space Habitat Baseball (Princeton)

On Earth, we can disregard the Coriolis force since it is only a second-order correction. If the player hits the ball with an initial velocity **v**, the maximum distance will be for the angle $\alpha = 45°$ (neglecting also the effects of air resistance). Then, we calculate the range L by decomposing the trajectory into its component motions, with initial velocities $v_x = v\cos\alpha$, $v_y = v\sin\alpha$

$$v\sin\alpha \cdot t - \frac{gt^2}{2} = 0, \qquad t = \frac{2v}{g}\sin\alpha$$

$$L = v\cos\alpha \cdot t = \frac{v^2\sin 2\alpha}{g}$$

resulting in an initial velocity off the bat of

$$v = \sqrt{gL/\sin 2\alpha} = \sqrt{gL} = \sqrt{10\cdot 120} \approx 34.6\,\text{m/s}. \qquad \text{(S.1.45.1)}$$

On the surface of the habitat we can no longer disregard the Coriolis force (see Figure S.1.45), so if we consider the problem in the rotating frame of the cylinder, the equations of motion become rather complicated. Therefore, let us view the exercise in the inertial frame. To provide the same apparent gravitational acceleration g, the cylinder has to rotate with an angular velocity $\omega_0 = \sqrt{g/R}$. The instantaneous linear velocity of the point P where the player stands will be

$$V = \omega_0 R = \sqrt{gR} = \sqrt{10\cdot 10^4} \approx 316\,\text{m/s} \qquad \text{(S.1.45.2)}$$

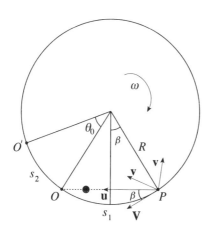

Figure **S.1.45**

When the ball leaves the surface, no forces act on it in the inertial frame. Its velocity is either u_+ or u_-, where the velocities of the ball in the rotating frame v_x and the habitat V are parallel or antiparallel, respectively:

$$u_+ = \sqrt{v_y^2 + (V + v_x)^2} = \sqrt{g\left(R + \sqrt{2LR} + L\right)}$$
$$u_- = \sqrt{v_y^2 + (V - v_x)^2} = \sqrt{g\left(R - \sqrt{2LR} + L\right)} \qquad \text{(S.1.45.3)}$$

The angle β of the line $\overline{PO}$ of length D to the tangent of the circle is found using (S.1.45.1)–(S.1.45.3):

$$\beta_\pm = \sin^{-1}\left(\frac{v_y}{u_\pm}\right) = \sin^{-1}\left(\frac{\sqrt{L/2}}{\sqrt{R \pm \sqrt{2LR} + L}}\right)$$

The distance along the surface of the cylinder PO will then be $s_1 = 2R\beta$. During the time of the flight, the cylinder rotates by an angle $\theta_\pm = \omega_0 t_\pm = \omega_0 D / u_\pm$, and the distance s_2 will be

$$s_2 = \theta_\pm R = \omega_0 \frac{D}{u_\pm} R = \sqrt{gR}\, \frac{2R \sin\beta_\pm}{u_\pm} \qquad \text{(S.1.45.4)}$$

The distance the player would hit the ball measured along the surface of the habitat is

$$L_\pm = |s_1 - s_2| = 2R\left|\beta_\pm - \frac{\sqrt{gR}}{u_\pm} \sin\beta_\pm\right|$$

$$= 2R\left|\beta_\pm - \frac{\sqrt{R}}{\sqrt{R \pm \sqrt{2LR} + L}} \sin\beta_\pm\right|$$

$$= 2R\left|\sin^{-1}\left(\frac{\sqrt{L/2}}{\sqrt{R \pm \sqrt{2LR} + L}}\right) - \frac{\sqrt{RL/2}}{R \pm \sqrt{2LR} + L}\right| \qquad \text{(S.1.45.5)}$$

Substituting numerical values into (S.1.45.5) we obtain

$$L_+ \approx 108\,\mathrm{m} \qquad L_- \approx 132\,\mathrm{m}$$

Therefore, to hit the furthest, the player should hit in the direction opposite to the direction of the habitat's rotation.

1.46   Vibrating String with Mass (Stony Brook)

a)   To derive the equation of motion of the string we assume that the oscillations of the string are small, the tension T is a slowly varying function

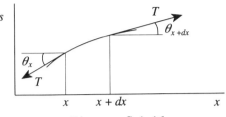

Figure S.1.46

of x, and there is no damping of the motion. Consider a part of the string between x and $x + dx$, where $s(x,t)$ is the transverse displacement of the string (see Figure S.1.46). The transverse force acting on this piece of mass ρdx is

$$\rho dx \cdot \frac{d^2 s(x,t)}{dt^2} = T \sin \theta|_{x+dx} - T \sin \theta|_x \qquad (S.1.46.1)$$

Using the initial assumptions, we can substitute a constant T for the tension $T(x)$ and write

$$T \sin \theta|_{x+dx} - T \sin \theta|_x = T(\sin \theta|_{x+dx} - \sin \theta|_x)$$

$$= T(\tan \theta|_{x+dx} - \tan \theta|_x) = T\left(\frac{ds}{dx}\bigg|_{x+dx} - \frac{ds}{dx}\bigg|_x\right) = T\frac{d^2 s}{dx^2}dx$$

$$(S.1.46.2)$$

where the substitution of $\tan \theta$ for $\sin \theta$ again follows from assumptions. Using (S.1.46.2) in (S.1.46.1), we obtain

$$\rho\frac{d^2 s(x,t)}{dt^2} = T\frac{d^2 s(x,t)}{dx^2} \qquad (S.1.46.3)$$

$$\ddot{s} = \frac{T}{\rho}s'' = c^2 s''$$

where $c = \sqrt{T/\rho}$ is the wave velocity.

b) Now, we have boundary conditions. We choose a standing wave solution. Another possible solution is a traveling wave $s(x,t) = f(x + ct) + g(x - ct)$, where f and g are some functions. In our case, we use the method of separation of variables: $s(x,t) = g(x)f(t)$, which, substituted into (S.1.46.3), gives

$$c^2\frac{g''}{g} = \frac{\ddot{f}}{f} = -\omega^2$$

where ω^2 is a constant independent of x and t. We arrive at two second-order differential equations for f and g and their solutions:

$$\ddot{f} + \omega^2 f = 0, \qquad f(t) = A\cos\omega t + B\sin\omega t$$

$$g'' + \frac{\omega^2}{c^2}\,g = 0, \quad g(x) = C\cos kx + D\sin kx, \quad k = \frac{\omega}{c}$$

Applying the boundary condition to the solution:

$$s(x,t)|_{x=0} = s(x,t)|_{x=L} = 0$$

from which we get

$$C = 0, \quad k = \frac{n\pi}{L}, \quad n = 0, \pm 1, \dots$$

For each mode $g_n(x) = \sin(\omega_n x/c)$ and $f_n(t) = A_n\cos\omega_n t + B_n\sin\omega_n t$ with $\omega_n \equiv nc\pi/L$. For each complete mode

$$s_n(x,t) = \sin(\omega_n x/c)(A_n\cos\omega_n t + B_n\sin\omega_n t)$$

and the transverse displacement is

$$s(x,t) = \sum_n s_n(x,t) = \sum_n \sin(n\pi x/L)[A_n\cos(nc\pi t/L) + B_n\sin(nc\pi t/L)]$$

c) To find the frequency change, use a perturbation method. Consider for simplicity the nth mode of the form $s_n = \sin(n\pi x/L)\cos\omega_n t$, corresponding to the initial conditions $s(x,0) = s_0(x)$ and $\dot{s}(x,0) = 0$. We know from the virial theorem that initially the average of potential energy of the string in the nth mode is equal to the kinetic energy:

$$\langle V_n \rangle = \langle K_n \rangle = \left\langle \frac{1}{2}\int_0^L \rho\dot{s}^2\,dx \right\rangle$$

$$= \left\langle \frac{1}{2}\int_0^L \rho\omega_n^2 \sin^2\left(\frac{n\pi x}{L}\right)\sin^2(\omega_n t)\,dx \right\rangle = \frac{\rho\omega_n^2 L}{8} \quad \text{(S.1.46.4)}$$

where we used $\langle \sin^2\omega_n t \rangle = \langle \cos^2\omega_n t \rangle = 1/2$. Now examine the nth mode of the string with mass to be of the same form as in (S.1.46.4): $s_n(x,t) = \sin(n\pi x/L)\cos\Omega_n t$ with a slightly different frequency Ω_n. Find the kinetic energy $\widetilde{K}_n$ in this mode of the string and then add the kinetic energy of the additional mass:

$$\widetilde{K}_n = \widetilde{K}_{\text{string}} + \widetilde{K}_{\text{mass}} = \frac{1}{2} \int_0^L \rho \dot{s}_n^2 \, dx + \frac{1}{2} \int_0^L m\delta\left(l - x\right) \dot{s}_n^2 \, dx \quad \text{(S.1.46.5)}$$

where $\delta\left(l - x\right)$ is the Dirac δ function. The average kinetic energy of the string with mass from (S.1.46.5)

$$\langle \widetilde{K}_n \rangle = \rho \frac{\Omega_n^2 L}{8} + \frac{1}{4} m\Omega_n^2 \sin^2\left(\frac{n\pi l}{L}\right) \quad \text{(S.1.46.6)}$$

where again we used $\langle \sin^2 \Omega_n t \rangle = \langle \cos^2 \Omega_n t \rangle = 1/2$. In this approximation, if we ignore the change in tension T, the average potential energy of the string with mass is the same as for the string alone, so $\langle V_n \rangle = \langle \widetilde{V}_n \rangle$. Utilizing this together with the virial theorem, which is also true for the modified string, we may write

$$\langle K_n \rangle = \langle V_n \rangle = \langle \widetilde{V}_n \rangle = \langle \widetilde{K}_n \rangle \quad \text{(S.1.46.7)}$$

So from (S.1.46.5)–(S.1.46.7)

$$\langle K_n \rangle - \langle \widetilde{K}_n \rangle_{\text{string}} = \langle \widetilde{K}_n \rangle_{\text{mass}}$$

or

$$\frac{1}{8}\rho\omega_n^2 L - \frac{1}{8}\rho\Omega_n^2 L = \frac{1}{4} m\Omega_n^2 \sin^2\left(\frac{\pi n l}{L}\right) \quad \text{(S.1.46.8)}$$

Therefore the new frequency

$$\Omega_n^2 \approx \frac{\omega_n^2}{1 + (2m/\rho L)\sin^2\left(n\pi l/L\right)}$$

$$\Omega_n \approx \omega_n \left(1 - \frac{m}{\rho L}\sin^2\frac{n\pi l}{L}\right) \quad \text{(S.1.46.9)}$$

where we used $m \ll \rho L$.

1.47   Shallow Water Waves (Princeton (a,b))

This problem is discussed in Landau & Lifshitz, *Fluid Mechanics*, Ch. 12. We essentially follow their solution. In this problem, we consider an incompressible fluid (which implies that the density ρ is constant (see Figure S.1.47). We also consider irrotational flow ($\nabla \times \mathbf{v} = 0$) and ignore the surface tension and viscosity of the fluid. This is a very idealized case; (Feynman calls it "dry" water in his *Lectures on Physics*). In this case,

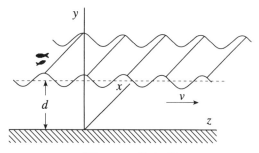

Figure S.1.47

we have $\nabla \cdot \rho \mathbf{v} = 0$ and since ρ is constant, $\nabla \cdot \mathbf{v} = 0$. Combining this equation with the condition $\nabla \times \mathbf{v} = 0$ allows us to introduce a potential φ (the so-called potential flow). The velocity $\mathbf{v}$ may be written in the form $\mathbf{v} = \nabla \varphi$, and for the potential we have

$$\nabla^2 \varphi = 0 \qquad (S.1.47.1)$$

On the bottom, we have the boundary condition

$$v_y = \left. \frac{\partial \varphi}{\partial y} \right|_{y=0} = 0 \qquad (S.1.47.2)$$

Using Euler's equation for an irrotational field

$$\frac{\partial \mathbf{v}}{\partial t} = -\nabla \left(\frac{p}{\rho} + \frac{v^2}{2} \right) + \mathbf{g} \qquad (S.1.47.3)$$

(Here p is pressure, $\mathbf{g} = -g\hat{\mathbf{y}}$ is the acceleration of gravity.) We substitute $\mathbf{v} = \nabla \varphi$ and rewrite (S.1.47.3) as

$$\nabla \left[\frac{\partial \varphi}{\partial t} + \frac{p}{\rho} + \frac{v^2}{2} + gy \right] = 0 \qquad (S.1.47.4)$$

Since (S.1.47.4) is the gradient of a function, the function itself will simply be

$$\frac{\partial \varphi}{\partial t} + \frac{p}{\rho} + \frac{v^2}{2} + gy = f(t)$$

where $f(t)$ is some arbitrary function of time which may be chosen to be zero. Also taking into account that $v^2/2 \ll gh$, we have

$$\frac{\partial \varphi}{\partial t} + \frac{p}{\rho} + gy = 0$$

or

$$p = -\rho g y - \rho \frac{\partial \varphi}{\partial t} \qquad (S.1.47.5)$$

Consider the surface of the unperturbed water at $y = d$ and introduce a small vertical displacement $Y = y - d$. Also, we assume that there is a constant pressure on the surface of the water p_0. Then from (S.1.47.5) we obtain

$$p_0 = -\rho g \left(Y + d\right) - \rho \frac{\partial \varphi}{\partial t} \qquad (\text{S.1.47.6})$$

The constant $p_0 + \rho g d$ can be eliminated by using another gauge for φ :

$$\varphi \to \varphi - \left(\frac{p_0}{\rho} + gd\right) t$$

We now obtain from (S.1.47.6)

$$Y = -\frac{1}{g} \frac{\partial \varphi}{\partial t}\bigg|_{y=Y+d} \qquad (\text{S.1.47.7})$$

Again using the fact that the amplitude of the waves is small, we can write $v_y = \partial Y/\partial t$. In the same approximation of small oscillations, we can take the derivative at $y = d$. On the other hand, $v_y = \partial \varphi/\partial y$. So, from (S.1.47.7)

$$\frac{\partial \varphi}{\partial y}\bigg|_{y=d} = \frac{\partial Y}{\partial t} = -\frac{1}{g} \frac{\partial^2 \varphi}{\partial t^2}\bigg|_{y=d} \qquad (\text{S.1.47.8})$$

Now look for a solution for φ in the form $\varphi = f(y)\cos(kz - \omega t)$. Substituting this into (S.1.47.1) gives

$$d^2 f/dy^2 - k^2 f = 0 \qquad (\text{S.1.47.9})$$

so

$$\varphi = \left(Ae^{ky} + Be^{-ky}\right)\cos(kz - \omega t) \qquad (\text{S.1.47.10})$$

where A, B are arbitrary constants. From (S.1.47.2), we find that $A = B$ and $\varphi = A' \cosh ky \cdot \cos(kz - \omega t)$ where $A' = 2A$. By differentiating the potential we obtain the velocity components

$$v_y = A'k \sinh ky \cdot \cos(kz - \omega t)$$

$$v_z = -A'k \cosh ky \cdot \sin(kz - \omega t)$$

b) From (S.1.47.8) we get the dispersion relation:

$$k \sinh kd = \frac{\omega^2}{g} \cosh kd \qquad (\text{S.1.47.11})$$

$$\omega^2 = gk \tanh kd \qquad (\text{S.1.47.12})$$

c) The group velocity of the waves is

$$u = \frac{\partial \omega}{\partial k} = \frac{\sqrt{g}}{2\sqrt{k \tanh kd}} \left(\tanh kd + \frac{kd}{\cosh^2 kd} \right) \quad \text{(S.1.47.13)}$$

Consider two limiting cases:

1) $kd \gg 1$, $d \gg \lambda$—short wavelength waves. Then

$$u \approx \frac{1}{2} \sqrt{g/k} = \frac{1}{2} \sqrt{g\lambda/2\pi}$$

2) $kd \ll 1$, $d \ll \lambda$—long wavelength waves. Then $u \approx \sqrt{gd}$.

1.48   Suspension Bridge (Stony Brook)

a) We use an elementary method to solve this problem. The conditions for a static equilibrium are

$$T(x + dx) \cdot \cos \theta(x + dx) - T(x) \cos \theta(x) = F_x = 0 \quad \text{(S.1.48.1)}$$

$$T(x + dx) \cdot \sin \theta(x + dx) - T(x) \sin \theta(x) = w \, dx \quad \text{(S.1.48.2)}$$

(see Figure S.1.48). (S.1.48.1) and (S.1.48.2) can be rewritten in the form

$$\frac{d}{dx} [T(x) \cos \theta(x)] = 0 \quad \text{(S.1.48.3)}$$

$$\frac{d}{dx} [T(x) \sin \theta(x)] = w \, dx \quad \text{(S.1.48.4)}$$

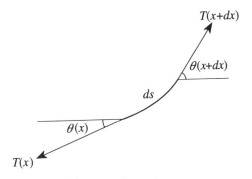

Figure **S.1.48**

Integrating (S.1.48.3) and (S.1.48.4), we obtain

$$T(x)\sin\theta(x) = C_0 + wx \qquad (S.1.48.5)$$

$$T(x)\cos\theta(x) = C_1 \qquad (S.1.48.6)$$

At $x = 0$, $\theta = 0$, so $C_0 = 0$; $C_1 = T_0$ and dividing (S.1.48.5) by (S.1.48.6), we have

$$\tan\theta(x) = \frac{dy}{dx} = \frac{w}{T_0}x \qquad (S.1.48.7)$$

From (S.1.48.7) we find the shape of the suspension bridge, which is parabolic

$$y = y_0 + \frac{wx^2}{2T_0} \qquad (S.1.48.8)$$

b)   To find the tension $T(x)$ at $x \neq 0$ ($\theta \neq 0$), multiply (S.1.48.5) by (S.1.48.6).

$$T^2(x)\sin\theta\cos\theta = T_0 wx \qquad (S.1.48.9)$$

$$T^2(x)\frac{\tan\theta}{1 + \tan^2\theta} = T_0 wx \qquad (S.1.48.10)$$

$$T^2(x) = T_0 wx\left(\tan\theta + \frac{1}{\tan\theta}\right) = T_0 wx\left(\frac{w}{T_0}x + \frac{T_0}{wx}\right) = T_0^2 + w^2 x^2$$

So

$$T(x) = T_0\sqrt{1 + (wx/T_0)^2} \qquad (S.1.48.11)$$

1.49   Catenary (Stony Brook, MIT)

a) Write the expressions for the length l and potential energy U (see Figure S.1.49) using

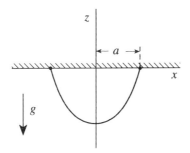

Figure **S.1.49**

$$dl^2 = dx^2 + dz^2$$

$$l = \int_{-a}^{a} \sqrt{1 + (dz/dx)^2}\, dx = \int_{-a}^{a} \sqrt{1 + z'^2}\, dx \qquad \text{(S.1.49.1)}$$

$$U = \int_{-a}^{a} \rho g z \sqrt{1 + z'^2}\, dx \qquad \text{(S.1.49.2)}$$

b) Here, we are not reproducing the usual Euler–Lagrange equations where we have minimized the action $I = \int \mathcal{L}\, dt$. Instead, we look for the minimum of U found in (a), subject to the constraint of constant length l. Utilizing the method of undetermined Lagrange multipliers, λ (see Goldstein, *Classical Mechanics*, Chapter 2), we may write

$$\mathcal{L} = \rho g z \sqrt{1 + z'^2} + \rho g \lambda \sqrt{1 + z'^2} \qquad \text{(S.1.49.3)}$$

The coefficient preceding λ simplifies the calculation. From (S.1.49.3)

$$\mathcal{L}^* = \sqrt{1 + z'^2}\,(z + \lambda)$$

where $\mathcal{L} = \rho g \mathcal{L}^*$

$$\frac{\partial \mathcal{L}^*}{\partial z'} = \frac{z'(z + \lambda)}{\sqrt{1 + z'^2}} \qquad\qquad \frac{\partial L^*}{\partial z} = \sqrt{1 + z'^2}$$

$$\frac{d}{dx}\left[\frac{z'(z + \lambda)}{\sqrt{1 + z'^2}}\right] - \sqrt{1 + z'^2} = 0 \qquad \text{(S.1.49.4)}$$

Before proceeding to (c), note that in this problem, we may immediately extract a first integral of the motion since $\mathcal{L}^*$ does not depend explicitly on x (see Goldstein, *Classical Mechanics* §2.6).

$$h \text{ (a constant)} \ = z'\frac{\partial \mathcal{L}^*}{\partial z'} - \mathcal{L}^*$$

$$= \frac{z'^2(z + \lambda)}{\sqrt{1 + z'^2}} - \sqrt{1 + z'^2}\,(z + \lambda) = -\frac{z + \lambda}{\sqrt{1 + z'^2}} \qquad \text{(S.1.49.5)}$$

c) We may now substitute $z = A\cosh(x/A) + B$ into (S.1.49.5), yielding

$$\sqrt{1 + z'^2} = \cosh(x/A)$$

$$h = -\frac{A\cosh(x/A) + B + \lambda}{\cosh(x/A)} \qquad \text{(S.1.49.6)}$$

h is constant for $\lambda = -B$. Calculate l from (S.1.49.1):

$$l = \int_{-a}^{a} \cosh(x/A)\, dx = 2A \sinh(a/A) \qquad \text{(S.1.49.7)}$$

Using (S.1.49.2) and (S.1.49.7), find U:

$$U = \rho g \int_{-a}^{a} [A \cosh(x/A) + B] \cosh(x/A)\, dx$$

$$= \rho g \frac{A}{2} \int_{-a}^{a} [\cosh(2x/A) + 1]\, dx + \rho g l B$$

$$= \frac{\rho g l A}{2} \cosh(a/A) + \rho g A a + \rho g l B \qquad \text{(S.1.49.8)}$$

Using $z(a) = z(-a) = 0$, we see that

$$B = -A \cosh(a/A)$$

(S.1.49.8) becomes

$$U = -\frac{\rho g l A}{2} \cosh(a/A) + \rho g A a \qquad \text{(S.1.49.9)}$$

From (S.1.49.7), we have

$$\cosh(a/A) = \sqrt{1 + (l/2A)^2}$$

so

$$U = -\frac{\rho g l A}{2} \sqrt{1 + (l/2A)^2} + \rho g A a = \rho g A a \left[1 - \frac{l}{2a} \sqrt{1 + (l/2A)^2} \right]$$

1.50   Rotating Hollow Hoop (Boston)

The Lagrangian for the system shown in Figure S.1.50 can be written in the form

$$\mathcal{L} = \frac{1}{2} m R^2 \dot{\theta}^2 + \frac{1}{2} m R^2 \dot{\varphi}^2 - \frac{1}{2} k R^2 (\varphi - \theta)^2$$

The generalized momenta are

$$p_\theta = m R^2 \dot{\theta} \qquad p_\varphi = m R^2 \dot{\varphi}$$

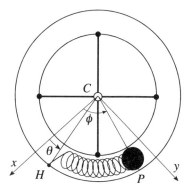

Figure **S.1.50**

The Hamiltonian is

$$\mathcal{H} = \frac{1}{2}\frac{p_\theta^2}{mR^2} + \frac{1}{2}\frac{p_\varphi^2}{mR^2} + \frac{1}{2}kR^2\left(\varphi - \theta\right)^2$$

Changing the variables gives

$$\varphi = \frac{\xi + \eta}{\sqrt{2}} \qquad\qquad \text{(S.1.50.1)}$$

$$\theta = \frac{\xi - \eta}{\sqrt{2}} \qquad\qquad \text{(S.1.50.2)}$$

$$\mathcal{L} = \frac{mR^2}{2}\left(\dot{\xi}^2 + \dot{\eta}^2\right) - kR^2\eta^2 \qquad\qquad \text{(S.1.50.3)}$$

Again, the generalized momenta are

$$p_\xi = mR^2\dot{\xi} \qquad p_\eta = mR^2\dot{\eta}$$

The Hamiltonian is

$$\mathcal{H} = p_\xi\dot{\xi} + p_\eta\dot{\eta} - \mathcal{L} = \frac{1}{2mR^2}\left(p_\xi^2 + p_\eta^2\right) + kR^2\eta^2$$

b) Since $\partial\mathcal{L}/\partial\xi = 0$ we have

$$p_\xi = \text{const} \qquad \dot{\xi} = \text{const}$$

The Poisson bracket of p_ξ with $\mathcal{H}$ is

$$[p_\xi, \mathcal{H}] = \frac{1}{2mR^2}\left(\left[p_\xi, p_\xi^2\right] + \left[p_\xi, p_\eta^2\right]\right) + kR^2\left[p_\xi, \eta^2\right] = 0$$

So, p_ξ is indeed the integral of motion since its Poisson bracket with $\mathcal{H}$ is equal to zero.

c) From (S.1.50.3), the equations of motion are

$$\ddot{\xi} = 0 \qquad\qquad (S.1.50.4)$$

$$mR^2\ddot{\eta} + 2kR^2\eta = 0 \qquad\qquad (S.1.50.5)$$

From (S.1.50.4) and (S.1.50.5) we obtain

$$\xi = \xi_0 + At$$

$$\eta = B\cos\omega t + C\sin\omega t, \qquad \omega = \sqrt{2k/m}$$

Using the initial conditions for (S.1.50.1) and (S.1.50.2) we have

$$\xi(0) = 0, \qquad \dot{\xi}(0) = 0$$

$$\eta(0) = \frac{\pi}{2\sqrt{2}}, \qquad \dot{\eta}(0) = 0$$

So

$$\xi \equiv 0$$

$$\eta = \frac{\pi}{2\sqrt{2}}\cos\omega t$$

and finally

$$\varphi = \frac{\xi+\eta}{\sqrt{2}} = \frac{\pi}{4}\cos\omega t$$

$$\theta = \frac{\xi-\eta}{\sqrt{2}} = -\frac{\pi}{4}\cos\omega t$$

1.51   Particle in Magnetic Field (Stony Brook)

a) A canonical transformation preserves the form of Hamilton's equations:

$$\dot{Q}_i = \frac{\partial\mathcal{H}}{\partial P_i} \qquad \dot{P}_i = -\frac{\partial\mathcal{H}}{\partial Q_i}$$

where $\mathcal{H} = \mathcal{H}(Q,P)$ is the transformed Hamiltonian. It can be shown that

Poisson brackets are invariant under such a transformation. In other words, for two functions f, g

$$[f, g]_{q,p} = [f, g]_{Q,P} \qquad (S.1.51.1)$$

where q, p and Q, P are the old and new variables, respectively. Since we have the following equations for Poisson brackets:

$$[q_i, q_j] = 0, \qquad [p_i, p_j] = 0, \qquad [q_i, p_j] = \delta_{ij} \qquad (S.1.51.2)$$

(S.1.51.1) and (S.1.51.2) combined give equivalent conditions for a transformation to be canonical:

$$[Q_i, Q_j]_{q,p} = 0, \qquad [P_i, P_j]_{q,p} = 0, \qquad [Q_i, P_j]_{q,p} = \delta_{ij}$$

Let us check for our transformation (we let $\alpha = 1/\sqrt{m\omega}$)

$$[X, P_x]_{Q,P} = \frac{\partial X}{\partial Q_1} \frac{\partial P_x}{\partial P_1} - \frac{\partial P_x}{\partial Q_1} \frac{\partial X}{\partial P_1} + \frac{\partial X}{\partial Q_2} \frac{\partial P_x}{\partial P_2} - \frac{\partial P_x}{\partial Q_2} \frac{\partial X}{\partial P_2}$$

$$= \alpha\sqrt{2P_1} \cos Q_1 \cdot \frac{1}{2\alpha} \frac{1}{\sqrt{2P_1}} \cos Q_1 - \left(-\frac{1}{2\alpha} \sqrt{2P_1} \sin Q_1 \right) \alpha \frac{\sin Q_1}{\sqrt{2P_1}}$$

$$-\alpha \left(-\frac{1}{2\alpha} \right)$$

$$= \frac{1}{2} + \frac{1}{2} = 1$$

and

$$[X, P_y]_{Q,P} = \frac{\partial X}{\partial Q_1} \frac{\partial P_y}{\partial P_1} - \frac{\partial P_y}{\partial Q_1} \frac{\partial X}{\partial P_1} + \frac{\partial X}{\partial Q_2} \frac{\partial P_y}{\partial P_2} - \frac{\partial P_y}{\partial Q_2} \frac{\partial X}{\partial P_2}$$

$$= \alpha\sqrt{2P_1} \cos Q_1 \left(-\frac{1}{2\alpha} \frac{1}{\sqrt{2P_1}} \sin Q_1 \right)$$

$$- \left(-\frac{\sqrt{2P_1}}{2\alpha} \cos Q_1 \right) \cdot \alpha \frac{\sin Q_1}{\sqrt{2P_1}}$$

$$= 0$$

Similarly

$$[Y, P_x]_{Q,P} = 0$$

$$[Y, P_y] = 1$$

and so on.

b)   For a particle in a magnetic field described by the vector potential $\mathbf{A} = (-YH/2, XH/2, 0)$, which corresponds to a constant magnetic field $\mathbf{H} = H\hat{\mathbf{z}}$, we should use the generalized momentum $\mathbf{P}$ in the Hamiltonian

$$p_x = P_x - \frac{e}{c}A_x$$

$$p_y = P_y - \frac{e}{c}A_y$$

so the Hamiltonian

$$\mathcal{H} = \frac{1}{2m}\left(P_x - \frac{e}{c}A_x\right)^2 + \frac{1}{2m}\left(P_y - \frac{e}{c}A_y\right)^2$$

$$= \frac{1}{2m}\left(P_x^2 + P_y^2\right) - \frac{e}{mc}\left(P_x A_x + P_y A_y\right) + \frac{e^2}{2mc^2}\left(A_x^2 + A_y^2\right)$$

$$= \frac{1}{2m}(P_x^2 + P_y^2) + \frac{eH}{2mc}(P_x Y - P_y X) + \frac{e^2 H^2}{8mc^2}(X^2 + Y^2)$$

$$= \frac{1}{2m}\frac{1}{4}m\omega(2P_1 - 2\sqrt{2P_1}Q_2\cos Q_1 + Q_2^2 - 2\sqrt{2P_1}P_2\sin Q_1 + P_2^2)$$

$$+ \frac{eH}{2mc}\frac{1}{2}(2P_1 - Q_2^2 - P_2^2)$$

$$+ \frac{e^2 H^2}{8mc^2}\frac{1}{m\omega}(2P_1 + 2\sqrt{2P_1}Q_2\cos Q_1 + Q_2^2 + 2\sqrt{2P_1}P_2\sin Q_1 + P_2^2)$$

$$= \omega P_1$$

So the Hamiltonian $\mathcal{H}$ does not depend on Q_1, Q_2, and P_2.

$$\dot{P}_1 = -\frac{\partial\mathcal{H}}{\partial Q_1} = 0$$

$$P_1 = \text{const} = \frac{E}{\omega}$$

$$\dot{Q}_1 = \frac{\partial\mathcal{H}}{\partial P_1} = \omega$$

$$Q_1 = \omega t + \alpha$$

where α is the initial phase. Also

$$P_2 = P_{20} = \sqrt{m\omega}X_0$$

$$Q_2 = Q_{20} = \sqrt{m\omega}Y_0$$

where X_0 and Y_0 are defined by the initial conditions. We can write this solution in terms of the variables X, Y, p_x, p_y :

$$X = \sqrt{2E/m\omega^2} \sin(\omega t + \alpha) + X_0$$

$$Y = \sqrt{2E/m\omega^2} \cos(\omega t + \alpha) + Y_0$$

$$p_x = P_x - \frac{e}{c}A_x = \sqrt{mE/2} \cos(\omega t + \alpha) - \frac{m\omega}{2}Y_0$$

$$+\frac{1}{2}\frac{eH}{c}\sqrt{2E/m\omega^2} \cos(\omega t + \alpha) + \frac{1}{2}\frac{eH}{c}Y_0$$

$$= \sqrt{2mE} \cos(\omega t + \alpha) = p_0 \cos(\omega t + \alpha)$$

Similarly

$$p_y = -p_0 \sin(\omega t + \alpha)$$

so this is indeed the solution for a particle moving in one plane in a constant magnetic field perpendicular to the plane.

1.52   Adiabatic Invariants (Boston (a)) and Dissolving Spring (Princeton, MIT (b))

a)

$$\frac{dI}{dt} = \frac{1}{2\pi} \oint_{C_t} \sum_i \dot{p}_i \, dq_i + p_i \, d\dot{q}_i = \frac{1}{2\pi} \oint_{C_t} \sum_i (\dot{p}_i \, dq_i - \dot{q}_i \, dp_i)$$

$$= \frac{1}{2\pi} \oint_{C_t} \sum_i \left(-\frac{\partial \mathcal{H}}{\partial q_i} \, dq_i - \frac{\partial \mathcal{H}}{\partial p_i} \, dp_i \right) = 0$$

For a harmonic oscillator $\mathcal{H}(p, q) = E$

$$\frac{p^2}{2m} + \frac{m\omega^2 q^2}{2} = E$$

This trajectory in phase space is obviously an ellipse:

$$\frac{p^2}{A^2} + \frac{q^2}{B^2} = 1$$

with

$$A = \sqrt{2mE}, \qquad B = \sqrt{2E/m\omega^2} \qquad\qquad (S.1.52.1)$$

The adiabatic invariant

$$I = \frac{1}{2\pi} \oint_C p \, dq = \frac{1}{2\pi} \int dp \, dq$$

where we transformed the first integral along the curve into phase area integral which is simply $I = \sigma/2\pi$, where σ is the area of an ellipse $\sigma = \pi AB$. So, taking A and B from (S.1.52.1) gives

$$I = \frac{AB}{2} = \frac{1}{2}\sqrt{2mE}\sqrt{2E/m\omega^2} = \frac{E}{\omega}$$

b) The fact that the spring constant decreases adiabatically implies that although the energy is not conserved its rate of change will be proportional to the rate of change in the spring constant: It can be shown (see for instance §49, Landau and Lifshitz, *Mechanics*) that in this approximation the quantity found in (a)—the so-called adiabatic invariant—remains constant. Our spring is of course a harmonic oscillator with frequency $\omega = \sqrt{K/m}$ and energy $E = (1/2)\,KA^2$. So we have

$$\frac{E_1}{\omega_1} = \frac{E_2}{\omega_2} \qquad\qquad (\text{S.1.52.2})$$

or

$$\sqrt{K_1}A_1^2 = \sqrt{K_2}A_2^2$$

So from (S.1.52.2), the new amplitude is

$$A_2 = \left(\frac{K_1}{K_2}\right)^{1/4} A_1$$

1.53 Superball in Weakening Gravitational Field (Michigan State)

The slow change of the acceleration of gravity implies that we will have an adiabatic invariant I as in Problem 1.52

$$I = \frac{1}{2\pi} \oint p \, dq$$

We have from energy conservation that

$$\frac{p^2}{2m} = \frac{p_0^2}{2m} - mgx$$

where x is the height of the ball or

$$p = \left(p_0^2 - 2m^2gx\right)^{1/2}$$

so

$$I = \frac{1}{2\pi} \oint_T p(x)\,dx = \frac{1}{\pi} \int_0^h p(x)\,dx = \frac{1}{\pi} \int_0^h \left(p_0^2 - 2m^2gx\right)^{1/2}\,dx$$

$$= -\frac{2}{3\pi} \left.\frac{\left(p_0^2 - 2m^2gx\right)^{3/2}}{2m^2g}\right|_0^h = \frac{p_0^3}{3\pi m^2 g}$$

where we used

$$\frac{p_0^2}{2m} = mgh$$

Therefore

$$\frac{p_0^3}{g} = \text{const} = \frac{p_1^3}{g_1}$$

and

$$v_1 = \left(\frac{g_1}{g}\right)^{1/3} v_0 = \left(\frac{0.9g}{g}\right)^{1/3} v_0 \approx 0.97 v_0$$

Relativity

2.1 Marking Sticks (Stony Brook)

a) According to observer O, the O' stick is Lorentz–contracted:

$$l' = l_0\sqrt{1-\beta^2} = 1 \cdot \sqrt{1-0.36} \text{ m} = 0.8 \text{ m}$$

where l_0 is the length of the stick in its rest frame. So observer O believes that the stick in O' is shorter. In this frame, the marking devices are triggered simultaneously at $t = 0$ when the origins of the two frames coincide (see Figure S.2.1a). As shown, the O stick will be marked at $x = 0.8$ m (in the O frame, the marking device of O' is at $x = 0.8$ m).

b) According to observer O', the O stick is Lorentz–contracted:

$$l = l_0\sqrt{1-\beta^2} = 1\sqrt{1-0.36} = 0.8 \text{ m}$$

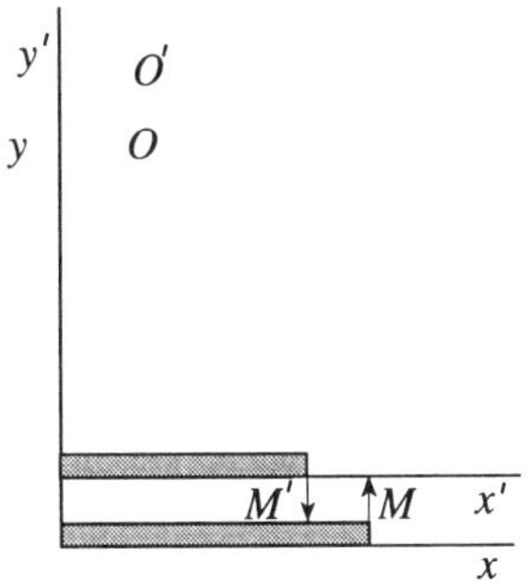

Figure S.2.1a

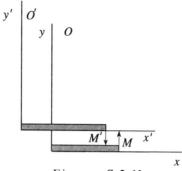

so that observer O' believes that the O stick is shorter. In frame O', the triggering of the devices (happening at $t = 0$ in frame O) is no longer simultaneous. Actually, the device in frame O' is triggered first, and that is why this observer agrees with the result found in (a) (see Figure S.2.1b). Indeed, from the Lorentz transformation of time, we can obtain the time in O' when his device is triggered:

$$0 = t = \gamma \left(t' + \frac{v}{c^2} x' \right)$$

So

$$t' = -\frac{v}{c^2} x'$$

The point in the O frame corresponding to the mark is

$$x = \gamma \left(x' + vt' \right) = \gamma x' \left(1 - \beta^2 \right) = \frac{x'}{\gamma} = 0.8 \text{ m}$$

the same result as in (a).

2.2   Rockets in Collision (Stony Brook)

a) In the Earth frame

$$t = \frac{l}{v_A + v_B} = \frac{b - a}{v_A + v_B} = \frac{4.2 \cdot 10^8 \text{ m}}{1.4 \cdot 3 \cdot 10^8 \text{ m/s}} = 1 \text{ s}$$

b) In A's frame, the coordinates of B are given by the Lorentz transformation from the Earth frame

$$\Delta x' = \gamma(\Delta x - v_A \Delta t)$$

$$\Delta t' = \gamma(\Delta t - \frac{v_A}{c^2} \Delta x)$$

So, in A's frame, B is approaching with velocity

$$v'_B = \frac{\Delta x'}{\Delta t'} = \frac{\gamma(\Delta x - v_A \Delta t)}{\gamma(\Delta t - v_A \Delta x/c^2)}$$

$$= \frac{v_B - v_A}{1 - v_A v_B/c^2} = \frac{-0.6c - 0.8c}{1 - (0.8c)(-0.6c)/c^2} = -0.95c$$

The same result (with opposite sign) may be obtained in B's frame for rocket A.

c) In each of the two rocket frames, the time to collision is dilated with respect to that in the Earth frame, and so

$$t_A = \frac{t}{\gamma_A} = 1 \cdot \sqrt{1 - 0.8^2} = 0.6 \text{ s}$$

$$t_B = \frac{t}{\gamma_B} = 1 \cdot \sqrt{1 - 0.6^2} = 0.8 \text{ s}$$

2.3   Photon Box (Stony Brook)

a) Consider the initial state of the system. Write the 4-momentum of the box and photons as p^μ_{box} and p^μ_{ph}, respectively:

$$p^\mu_{box} = (M_0 c, 0)$$

$$p^\mu_{ph} = \left(\frac{N h \nu_0}{c}, 0 \right)$$

(S.2.3.1)

where we have used the fact that for a standing wave (which can be represented as the sum of traveling waves with opposite momenta) the total momentum is zero. Therefore, from (S.2.3.1), the 4-momentum of the system P^μ is given by

$$P^\mu = p^\mu_{box} + p^\mu_{ph} = \left(M_0 c + \frac{N h \nu_0}{c}, 0 \right)$$

(S.2.3.2)

From (S.2.3.2), we can find the rest mass of the total system M (which is defined by $M^2 c^2 = P^\mu P_\mu$)

$$M = M_0 + \frac{N h \nu_0}{c^2}$$

(S.2.3.3)

b) Transform the 4-momentum by going into a frame moving with velocity $-v$ along the x axis. We have in this frame for energy $\mathcal{E}'$ and momentum P'

$$\mathcal{E}' = \gamma \left(\mathcal{E} + vP\right) = \gamma \mathcal{E}$$

$$P' = \gamma \left(P + \beta \frac{\mathcal{E}}{c}\right) = \gamma \beta \frac{\mathcal{E}}{c}$$

where $\mathcal{E}$ and p are the total energy and momentum in the rest frame, respectively. So

$$P'^{\mu} = \left(\gamma \frac{\mathcal{E}}{c}, \gamma \beta \frac{\mathcal{E}}{c}\right)$$

Therefore in the moving frame

$$M'^2 c^2 = \gamma^2 \frac{\mathcal{E}^2}{c^2} - \gamma^2 \beta^2 \frac{\mathcal{E}^2}{c^2} = \gamma^2 \frac{\mathcal{E}^2}{c^2} \left(1 - \beta^2\right) = \frac{\mathcal{E}^2}{c^2}$$

and

$$M' = \frac{\mathcal{E}}{c^2} = M = M_0 + \frac{N h \nu_0}{c^2}$$

We expect this to be true, of course, since mass is a relativistic invariant under a Lorentz transformation.

Another way to look at it is to consider a transformation of energy and momentum of the photons and the box separately. The frequencies of these photons will be Doppler shifted (see Problem 2.12):

$$\nu_1 = \nu \sqrt{\frac{1 + \beta}{1 - \beta}}$$

$$\nu_2 = \nu \sqrt{\frac{1 - \beta}{1 + \beta}}$$

The energy of the photons

$$\mathcal{E}'_{ph} = \frac{N h}{2} \left(\nu_1 + \nu_2\right) = \frac{N h \nu_0}{\sqrt{1 - \beta^2}} = \gamma N h \nu_0$$

The energy of the box $\mathcal{E}'_{\text{box}} = \gamma M_0 c^2$. The momentum of the photons

$$p'_{ph} = \frac{N h}{2c} \left(\nu_1 - \nu_2\right) = \frac{N h \nu_0}{2c} \frac{2\beta}{\sqrt{1 - \beta^2}} = \gamma \beta N h \nu_0$$

The momentum of the box

$$p'_{\text{box}} = \gamma M_0 v$$

So the 4-momentum is the same as found above

$$P'^{\mu} = \left(\gamma M_0 c + \gamma \frac{N h \nu_0}{c}, \gamma M_0 v + \gamma \beta \frac{N h \nu_0}{c}\right) = \left(\gamma \frac{\mathcal{E}}{c}, \gamma \beta \frac{\mathcal{E}}{c}\right)$$

2.4 Cube's Apparent Rotation (Stanford, Moscow Phys-Tech)

At any given moment, the image of the cube is created by the photons reaching the observer at this time. The light received from points A and B of the near face of the cube is accompanied by light from point D emitted a time $(1/c)$ earlier (see Figure S.2.4a). The length of $\overline{AB}$ is Lorentz contracted to $\sqrt{1 - \beta^2}$, while the distance from A' to A is β (the distance the

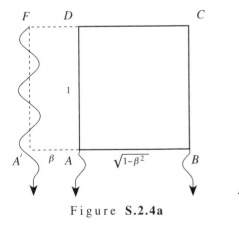

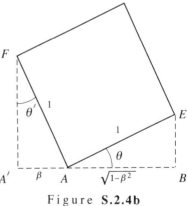

Figure **S.2.4a**   Figure **S.2.4b**

cube has moved while the light from D travelled to the front face). The apparent rotation is seen in Figure S.2.4b. The angle of rotation θ, $\angle BAE$, should be equal to θ', $\angle A'FA$. From the figure, we see that $\cos\theta = \sqrt{1 - \beta^2}$ and $\sin\theta' = \beta$. So $\theta = \theta'$, and the cube does appear rotated by $\sin^{-1}\beta$. A more detailed solution of this problem employs the Lorentz transformation from frame K' to K of the velocities which leads to the light aberration seen by the observer O [see *Phys. Rev.* **116**, 1041 (1959)].

2.5   Relativistic Rocket (Rutgers)

a) Let us consider the short interval dt in the center of mass frame moving with velocity v; the fuel is ejected with velocity u in this frame. At time $t' = t + dt$ the velocity of the rocket increases by dv. The mass $M(t)$ of the rocket decreases by dM ($dM < 0$) and the mass $|dM|$ of the ejected fuel will have a velocity $-u$ in this frame. Momentum conservation gives

$$M(t + dt)\, dv - u\, |dM| = M(t + dt)dv + udM = 0 \qquad (S.2.5.1)$$

where $M(t+dt)$ is the mass of the rocket at time $t+dt$. Expanding $M(t+dt)$ as

$$M(t + dt) = M(t) + M'(t)\, dt$$

and neglecting second-order terms in the differentials yields

$$M\, dv = -u\, dM \qquad (S.2.5.2)$$

Transforming to the lab frame and using $dv = dV$, where V is the velocity of the rocket in the lab frame, we obtain a solution for the initial condition $v(0) = 0$:

$$\frac{M}{M_0} = e^{-V/u} \qquad (S.2.5.3)$$

b) Write down momentum conservation in the rocket's frame:

$$M\, dv - \gamma_u\, dm\, u = 0 \qquad (S.2.5.4)$$

where M is the mass of the rocket, dm is the mass of the fuel, and $\gamma_u = 1/\sqrt{1 - u^2/c^2}$. Energy conservation in the frame of the rocket gives

$$M = \gamma_u\, dm + (M + dM) \qquad (S.2.5.5)$$

We ignored the relativistic corrections to the mass of the rocket in (S.2.5.5) and terms such as $1/\sqrt{1 - (dv)^2/c^2}$ in (S.2.5.4). Substituting dm from (S.2.5.5) into (S.2.5.4), we have

$$M\, dv = -u\, dM \qquad (S.2.5.6)$$

which is the same result as obtained in the nonrelativistic calculation of (a). Now we must transform dv from the instantaneous rocket frame to the laboratory frame. Using the equation for the addition of velocities, we have

$$V + dV = \frac{V + dv}{1 + V\, dv/c^2} \qquad (S.2.5.7)$$

where $V+dV$ is the new velocity of the rocket in the lab frame. Rearranging

(S.2.5.7) gives

$$dV + \frac{V^2}{c^2}\, dv = dv$$

$$\text{(S.2.5.8)}$$

$$dv = \gamma^2\, dV$$

where $\gamma = 1/\sqrt{1 - V^2/c^2}$. Substituting (S.2.5.8) into (S.2.5.6), we obtain $M\gamma^2\, dV = -u\, dM$ or

$$\frac{dV}{1 - V^2/c^2} = \frac{1}{2}\left(\frac{dV}{1 - V/c} + \frac{dV}{1 + V/c}\right) = -u\frac{dM}{M} \qquad \text{(S.2.5.9)}$$

Integrating (S.2.5.9), we have

$$\frac{c}{2}\left[-\ln(1 - \beta) + \ln(1 + \beta)\right] = u\ln\frac{M_0}{M} \qquad \text{(S.2.5.10)}$$

where $\beta = V/c$, from which we find

$$\frac{M}{M_0} = \left(\frac{1 - \beta}{1 + \beta}\right)^{c/2u} \qquad \text{(S.2.5.11)}$$

If $\beta \ll 1$ then (S.2.5.11) boils down to

$$\frac{M}{M_0} \approx e^{-V/2u} \cdot e^{-V/2u} = e^{-V/u}$$

the same result as that obtained in (a).

2.6 Rapidity (Moscow Phys-Tech)

a) The velocity of the particle moving in frame K_2 with velocity v_2 in the frame K_1 is given by a standard formula:

$$v_1 = \frac{v_2 + v}{1 + v_2 v/c^2}$$

Introducing $\beta_i = v_i/c$, we may rewrite this formula in the form

$$\beta_1 = \frac{\beta_2 + \beta}{1 + \beta_2\beta} \qquad \text{(S.2.6.1)}$$

Now the same formula may be written for a transformation from K_1 to

K_0 :

$$\beta_0 = \frac{\beta_1 + \beta}{1 + \beta_1 \beta} \qquad (S.2.6.2)$$

Now substituting (S.2.6.1) into (S.2.6.2), we obtain

$$\beta_0 = \frac{\beta_2 + 2\beta/(1 + \beta^2)}{1 + 2\beta_2\beta/(1 + \beta^2)}$$

b) If we need to make a transformation for n-frames, it is difficult to obtain a formula using the approach in (a). Instead, we use the idea of rapidity, ψ. Indeed for one frame, we had in (S.2.6.1)

$$\beta_1 = \frac{\beta_2 + \beta}{1 + \beta_2 \beta}$$

which is the formula for the tanh of a sum of arguments

$$\tanh(\psi_1 + \psi_2) = \frac{\tanh\psi_1 + \tanh\psi_2}{1 + \tanh\psi_1 \tanh\psi_2}$$

where $\tanh\psi_i = \beta_i$. This means that the consecutive Lorentz transformations are equivalent to adding rapidities. So the velocity in the frame K_0 after n transformations (if $v_{n+1} = v$) will be given by

$$\beta_0 = \tanh\psi = \tanh\sum_i \psi_i = \tanh[(n+1)\tanh^{-1}\beta]$$

We can check that if $n \to \infty$, then $\beta_0 \to 1$.

2.7 Charge in Uniform Electric Field (Stony Brook, Maryland, Colorado)

The plane of motion of a particle will be defined by its initial velocity $\mathbf{v}$ and the direction of the electric field $\mathbf{E}$. Let the initial velocity coincide with the x axis and $\mathbf{E}$ with the y axis. We may write the equations of motion for a charge in an electric field

$$\frac{d\mathbf{p}}{dt} = e\mathbf{E} \qquad (S.2.7.1)$$

where $\mathbf{p}$ is the momentum of the particle. Obviously, since there is no force in the direction perpendicular to the x–y plane, the particle will move in this plane at all later times. We can write (S.2.7.1) in the form

$$\frac{dp_x}{dt} = 0 \qquad (S.2.7.2)$$

$$\frac{dp_y}{dt} = eE \qquad (S.2.7.3)$$

Integrating (S.2.7.2) and (S.2.7.3) yields

$$p_x = p_{x_0} = p_0 \qquad (S.2.7.4)$$

$$p_y = eEt \qquad (S.2.7.5)$$

The energy $\mathcal{E}$ of the particle (without the potential energy due to the field) is given by

$$\mathcal{E} = \sqrt{m^2 c^4 + p^2 c^2} = \sqrt{m^2 c^4 + p_0^2 c^2 + c^2 e^2 E^2 t^2}$$

$$= \sqrt{\mathcal{E}_0^2 + (ceEt)^2} \qquad (S.2.7.6)$$

where $\mathcal{E}_0 = \sqrt{m^2 c^4 + p_0^2 c^2}$ is the initial energy of the particle. The work done by the electric field changes the energy of the particle

$$\frac{d\mathcal{E}}{dt} = e\mathbf{E} \cdot \mathbf{v} = eEv_y = eE\frac{dy}{dt} \qquad (S.2.7.7)$$

or

$$\mathcal{E} = \mathcal{E}_0 + eEy \qquad (S.2.7.8)$$

Equations (S.2.7.6) and (S.2.7.8) result in

$$\mathcal{E}_0 + eEy = \sqrt{\mathcal{E}_0^2 + (ceEt)^2} \qquad (S.2.7.9)$$

which yields

$$y = \frac{\mathcal{E}_0}{eE}\left\{ \sqrt{1 + (ceEt)^2 / \mathcal{E}_0^2} - 1 \right\} \qquad (S.2.7.10)$$

and

$$t = \frac{\sqrt{(\mathcal{E}_0 + eEy)^2 - \mathcal{E}_0^2}}{ceE} \qquad (S.2.7.11)$$

On the other hand

$$\frac{p_y}{p_x} = \frac{\gamma m v_y}{\gamma m v_x} = \frac{v_y}{v_x} = \frac{dy/dt}{dx/dt} = \frac{dy}{dx} \qquad (S.2.7.12)$$

Substituting $p_x = p_0$ and $p_y = eEt$ into (S.2.7.12) and using t from (S.2.7.11), we find

$$\frac{dy}{dx} = \frac{eEt}{p_0} = \frac{\sqrt{(\mathcal{E}_0 + eEy)^2 - \mathcal{E}_0^2}}{p_0 c} \qquad (S.2.7.13)$$

Integrating (S.2.7.13), we obtain

$$\frac{x}{p_0 c} = \int \frac{dy}{\sqrt{(\mathcal{E}_0 + eEy)^2 - \mathcal{E}_0^2}} = \frac{1}{eE} \cosh^{-1} \frac{eEy}{\mathcal{E}_0} + \text{const}$$

For the initial conditions $x_0 = y_0 = 0$

$$y = \frac{\mathcal{E}_0}{eE} \left(\cosh \frac{eEx}{p_0 c} - 1 \right) \qquad (S.2.7.14)$$

So the particle in a constant electric field moves along a catenary (see Figure S.2.7, where we took $e > 0$). If the velocity of the particle $v \ll c$, then $p_0 = mv_0$, $\mathcal{E}_0 = mc^2$, and expanding $\cosh(eEx/p_0 c)$, we obtain

$$y \approx \frac{eE}{2mv_0^2} x^2$$

which gives the classical result for a charged particle in an electric field. Also note that (S.2.7.10) coincides with the result for uniformly accelerated motion in the proper reference frame, where the acceleration $w_0 = eE/m$ and $p_0 = 0$ (see Problem 2.9, (S.2.9.7)). Under Lorentz transformations for frames moving with velocities parallel to the electric field E, the field is unchanged (see Landau and Lifshitz, *The Classical Theory of Fields*, Chapter 3).

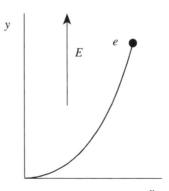

Figure   S.2.7

2.8   Charge in Electric Field and Flashing Satellites (Maryland)

Starting from a 4-vector potential $(\phi, \mathbf{A})$, we can obtain equations of motion for a charged particle in the electromagnetic field

$$\frac{d\mathbf{p}}{dt} = -\frac{q}{c}\frac{\partial \mathbf{A}}{\partial t} - q\nabla\phi - \frac{q}{c}\mathbf{v} \times (\nabla \times \mathbf{A}) \qquad \text{(S.2.8.1)}$$

By definition

$$\mathbf{E} = -\frac{1}{c}\frac{\partial \mathbf{A}}{\partial t} - \nabla\phi$$

$$\mathbf{H} = \nabla \times \mathbf{A}$$

Therefore (S.2.8.1) becomes

$$\frac{d\mathbf{p}}{dt} = q\mathbf{E} + \frac{q}{c}\mathbf{v} \times \mathbf{H} \qquad \text{(S.2.8.2)}$$

In this case of one-dimensional motion, where there is only an electric field E and momentum p in the x direction, we obtain

$$\frac{dp}{dt} = qE \qquad \text{(S.2.8.3)}$$

where

$$p = \frac{mv}{\sqrt{1-\beta^2}} = \frac{mc\beta}{\sqrt{1-\beta^2}}, \qquad \beta = v/c \qquad \text{(S.2.8.4)}$$

b) To show that (P.2.8.1) is a solution to (S.2.8.3), we write

$$\beta c = \frac{dx}{dt} = \frac{dx/d\tau}{dt/d\tau}$$

Now

$$\frac{dx}{d\tau} = c\sinh(qE\tau/mc) \qquad \frac{dt}{d\tau} = \cosh(qE\tau/mc)$$

so

$$\beta c = c\tanh(qE\tau/mc)$$

Since $1 - \tanh^2 = 1/\cosh^2$, we may rewrite (S.2.8.4)

$$p = \frac{mc\tanh(qE\tau/mc)}{1/\cosh(qE\tau/mc)} = mc\sinh(qE\tau/mc)$$

Differentiating,

$$\frac{dp}{dt} = \frac{dp/d\tau}{dt/d\tau} = \frac{qE\cosh(qE\tau/mc)}{\cosh(qE\tau/mc)} = qE$$

verifying (S.2.8.3). To show that τ is the proper time for the particle, we must demonstrate that

$$c^2 d\tau^2 = c^2 dt^2 - dx^2$$

From (P.2.8.1)

$$dx = c\sinh(qE\tau/mc)\, d\tau \qquad dt = \cosh(qE\tau/mc)\, d\tau$$

So

$$c^2 dt^2 - dx^2 = c^2(\cosh^2 - \sinh^2)\, d\tau^2 = c^2\, d\tau^2$$

as required.

c) Define the 4-momentum as $(\mathcal{E}/c, p, 0, 0)$, where $\mathcal{E}$ is the energy $\mathcal{E} = mc^2\gamma$ and p is the momentum $p = mc\beta\gamma$. The 4-acceleration is given by $(1/m)(dp^\mu/d\tau)$

$$w^\mu = \left(\frac{qE}{m}\sinh(qE\tau/mc), \frac{qE}{m}\cosh(qE\tau/mc), 0, 0\right)$$

$$w^\mu w_\mu = \left(\frac{qE}{m}\right)^2 (\sinh^2 - \cosh^2) = -\left(\frac{qE}{m}\right)^2$$

From (P.2.8.1), $x^2 - c^2t^2 = (mc^2/qE)^2$, which defines a hyperbola (see Figure S.2.8a).

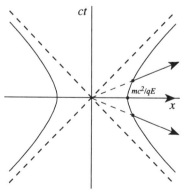

Figure S.2.8a

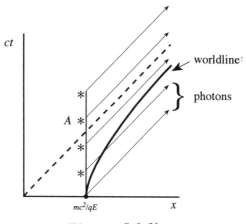

Figure **S.2.8b**

d) From Figure S.2.8b, we see that flashes emitted at a constant frequency f will cross the worldline of the particle until point A, where the trajectory of the satellite is above $ct = x$. To find the number of flashes, we find the time of intersection of $x = mc^2/qE$ and $x = ct$, i.e., $t = mc/qE$. The number of flashes is therefore $ft = mcf/qE$.

e) As shown in Figure S.2.8c, we need the intersection of

$$ct = -x + b$$

and

$$x^2 - c^2t^2 = \left(\frac{mc^2}{qE}\right)^2 = x_0^2$$

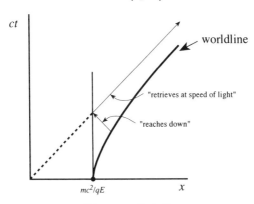

Figure **S.2.8c**

where b is found from

$$\frac{mc^2}{qE} = -\frac{mc^2}{qE} + b$$

$$b = \frac{2mc^2}{qE}$$

Thus,

$$x = -ct + \frac{2mc^2}{qE}$$

$$x^2 = c^2t^2 - \frac{4mc^3t}{qE} + \frac{4m^2c^4}{q^2E^2}$$

Therefore,

$$x^2 - c^2t^2 = \frac{m^2c^4}{q^2E^2} = -\frac{4mc^3t}{qE} + \frac{4m^2c^4}{q^2E^2}$$

and

$$t = \frac{3}{4}\frac{mc}{qE}$$

2.9   Uniformly Accelerated Motion (Stony Brook)

a) The 4-velocity, by definition

$$u^\mu = \frac{dx^\mu}{d\tau} = \frac{dx^\mu}{dt}\frac{dt}{d\tau}$$

where $d\tau = dt\sqrt{1 - v^2/c^2}$. Therefore,

$$u^\mu = \frac{1}{\sqrt{1 - v^2/c^2}}\frac{dx^\mu}{dt} = \begin{cases} \dfrac{c}{\sqrt{1 - v^2/c^2}}, & \mu = 0 \\[3mm] \dfrac{v^\mu}{\sqrt{1 - v^2/c^2}}, & \mu = 1, 2, 3 \end{cases}$$

or

$$u^\mu = (c\gamma, \mathbf{v}\gamma)$$

b) For arbitrary velocity $\mathbf{v}$, the 4-acceleration

$$w^\mu = \frac{d^2 x^\mu}{d\tau^2} = \frac{du^\mu}{d\tau} = \frac{du^\mu}{dt}\frac{dt}{d\tau} = \frac{1}{\sqrt{1-\beta^2}}\frac{du^\mu}{dt}$$

For $\mu = 0$

$$\frac{du^0}{dt} = \frac{\mathbf{v}\cdot\dot{\mathbf{v}}}{c(1-\beta^2)^{3/2}}$$

For $\mu = 1,2,3$

$$\frac{du^\mu}{dt} = \frac{\dot{v}^\mu}{\sqrt{1-\beta^2}} + \frac{v^\mu}{c^2}\frac{\mathbf{v}\cdot\dot{\mathbf{v}}}{(1-\beta^2)^{3/2}}$$

Therefore, we find

$$w^\mu = \left(\frac{1}{c}\frac{\mathbf{v}\cdot\dot{\mathbf{v}}}{(1-\beta^2)^2}, \frac{\dot{\mathbf{v}}}{(1-\beta^2)} + \frac{\mathbf{v}(\mathbf{v}\cdot\dot{\mathbf{v}})}{c^2(1-\beta^2)^2}\right) \qquad (S.2.9.1)$$

In the proper frame of reference, where the velocity of the particle $\mathbf{v} = 0$ at any given moment, and assuming $\dot{\mathbf{v}} = \dot{v}_x = w_0 = \text{const}$,

$$w^\mu = (0, w_0, 0, 0)$$

Using $w^2 = w^\mu w_\mu$, we have

$$w^2 = -w_0^2 = \text{const}$$

c) $w^\mu w_\mu$ from (S.2.9.1) may be written in the form

$$w^\mu w_\mu = \frac{1}{c^2}\frac{(\mathbf{v}\cdot\dot{\mathbf{v}})^2}{(1-\beta^2)^4} - \frac{\dot{v}^2}{(1-\beta^2)^2} - \frac{v^2(\mathbf{v}\cdot\dot{\mathbf{v}})^2}{c^4(1-\beta^2)^4} - \frac{2(\mathbf{v}\cdot\dot{\mathbf{v}})^2}{c^2(1-\beta^2)^3}$$

$$= \frac{\gamma^8}{c^2}(\mathbf{v}\cdot\dot{\mathbf{v}})^2\left[1-\beta^2-2(1-\beta^2)\right] - \gamma^4\dot{v}^2$$

$$= -\frac{\gamma^6(\mathbf{v}\cdot\dot{\mathbf{v}})^2}{c^2} - \gamma^4\dot{v}^2 \qquad (S.2.9.2)$$

Using the identity $(\mathbf{A}\times\mathbf{B})^2 = A^2 B^2 - (\mathbf{A}\cdot\mathbf{B})^2$, we may rewrite (S.2.9.2)

$$w^\mu w_\mu = -\gamma^6\left(\dot{v}^2 - (\dot{\mathbf{v}}\times\mathbf{v}/c)^2\right) \qquad (S.2.9.3)$$

In the fixed frame, since the acceleration is parallel to the velocity, (S.2.9.3) reduces to

$$w^\mu w_\mu = -\gamma^6\dot{v}^2$$

So, given that w^2 is a relativistic invariant, $w_0 = \gamma^3 \dot{v}$. Now, differentiating

$$\frac{d}{dt} \frac{v}{\sqrt{1-\beta^2}} = \frac{\dot{v}}{\sqrt{1-\beta^2}} + \frac{\dot{v}\beta^2}{(1-\beta^2)^{3/2}}$$

$$= \frac{\dot{v}(1-\beta^2) + \dot{v}\beta^2}{(1-\beta^2)^{3/2}} = \frac{\dot{v}}{(1-\beta^2)^{3/2}} = \gamma^3 \dot{v} = w_0 \qquad \text{(S.2.9.4)}$$

d) Integrating (S.2.9.4), we have

$$\frac{v}{\sqrt{1-\beta^2}} = w_0 t + \text{const} \qquad \text{(S.2.9.5)}$$

Taking $v = 0$ at $t = 0$, we obtain

$$\frac{v^2}{1-v^2/c^2} = w_0^2 t^2$$

and so

$$v = \frac{w_0 t}{\sqrt{1 + w_0^2 t^2/c^2}} \qquad \text{(S.2.9.6)}$$

As $w_0 t \to \infty$, $v \to c$. Integrating (S.2.9.6) with $x(0) = 0$ yields

$$x = \frac{c^2}{w_0} \left\{ \sqrt{1 + w_0^2 t^2/c^2} - 1 \right\} \qquad \text{(S.2.9.7)}$$

As $c \to \infty$ (classical limit), (S.2.9.6) and (S.2.9.7) become

$$v = w_0 t \qquad x = \frac{w_0 t^2}{2}$$

appropriate behavior for a uniformly accelerated classical particle.

2.10 Compton Scattering (Stony Brook, Michigan State)

a) From momentum and energy conservation we can write

$$\mathbf{p} = \tilde{\mathbf{p}} + \mathbf{p}_e$$
$$\mathcal{E} + \mathcal{E}_e = \tilde{\mathcal{E}} + \tilde{\mathcal{E}}_e \qquad \text{(S.2.10.1)}$$

where $\mathbf{p}, \tilde{\mathbf{p}}, \mathcal{E}, \tilde{\mathcal{E}}$ are the momenta and energies of the photon before and after the scattering, respectively, $\mathbf{p}_e$, $\tilde{\mathcal{E}}_e$ are the final momentum and energies of the electron, and $\mathcal{E}_e$ is its initial energy. We have for the electron

$$\tilde{\mathcal{E}}_e = \sqrt{p_e^2 c^2 + m^2 c^4}, \qquad \mathcal{E}_e = mc^2$$

and for the photon
$$\mathcal{E} = pc \qquad \tilde{\mathcal{E}} = \tilde{p}c$$

So we can rewrite (S.2.10.1) in the form

$$\mathbf{p} - \tilde{\mathbf{p}} = \mathbf{p}_e$$

$$pc + mc^2 = \tilde{p}c + \sqrt{p_e^2 c^2 + m^2 c^4}$$

(S.2.10.2)

b) To solve these equations we can express the momentum of the recoil electron $\mathbf{p}_e$ in two ways

$$\mathbf{p}_e^2 = (\mathbf{p} - \tilde{\mathbf{p}})^2$$

$$\mathbf{p}_e^2 = (p - \tilde{p})^2 + 2mc(p - \tilde{p})$$

(S.2.10.3)

from (S.2.10.2).
$$p\tilde{p}(1 - \cos\theta) = mc(p - \tilde{p})$$

and for a special case $\theta = \pi/2$, $\cos\theta = 0$. We have

$$p\tilde{p} = mc(p - \tilde{p})$$

Dividing this equation by $p\tilde{p}$ we get

$$1 = mc\left(\frac{1}{\tilde{p}} - \frac{1}{p}\right)$$

Taking into account that $p = h/\lambda$, we obtain the final result:

$$\tilde{\lambda} - \lambda = \frac{h}{mc}$$

2.11 Mössbauer Effect (Moscow Phys-Tech, MIT, Colorado)

Write the energy and momentum conservation equations:

$$p + p_{ph} = 0$$

(S.2.11.1)

$$mc^2 + \Delta\mathcal{E} = \hbar\omega + \sqrt{p^2c^2 + m^2c^4} \qquad (S.2.11.2)$$

where p is the momentum of the atom after emitting the photon, $p_{ph} = \hbar\omega/c$, the momentum of the photon, and ω is the photon's frequency. Substituting $p = -\hbar\omega/c$ from (S.2.11.1) into (S.2.11.2) and rewriting it in the form $mc^2 + \Delta\mathcal{E} - \hbar\omega = \sqrt{\hbar^2\omega^2 + m^2c^4}$, we can find, after squaring both sides,

$$\omega = \frac{\Delta\mathcal{E}}{2\hbar} \left(\frac{\Delta\mathcal{E} + 2mc^2}{\Delta\mathcal{E} + mc^2} \right) \qquad (S.2.11.3)$$

Now taking into account that $\Delta\mathcal{E} + mc^2 = Mc^2$, we can rewrite (S.2.11.3) as

$$\omega = \frac{\Delta\mathcal{E}}{\hbar} \left(1 - \frac{\Delta\mathcal{E}}{2Mc^2} \right) \qquad (S.2.11.4)$$

which is smaller by the amount of $(\Delta\mathcal{E})^2/2Mc^2\hbar$ than it would have been without relativistic effects. In the case of a crystalline lattice (Mössbauer effect), the atoms are strongly coupled to the lattice and have an effective mass $M_0 >> M$. From equation (S.2.11.4) we can see that in this case the atom practically does not absorb energy, which all goes into the energy of the photon, and therefore there is no frequency shift due to this effect.

2.12 Positronium and Relativistic Doppler Effect (Stony Brook)

a) The two photons will have momenta p of the same magnitude (but opposite sign), so their energies $\mathcal{E}$ are also the same:

$$2\mathcal{E} = 2m_ec^2 + \mathcal{E}_b$$

$$\mathcal{E} = pc$$

where we have chosen their momenta to be along the x axis. From these formulas we can find

$$\mathcal{E} = m_ec^2 + \frac{\mathcal{E}_b}{2} \qquad (S.2.12.1)$$

$$p_\pm = \pm\frac{\mathcal{E}}{c} = \pm \left(m_ec + \frac{\mathcal{E}_b}{2c} \right) \qquad (S.2.12.2)$$

The frequency of the photons is

$$\omega = \frac{\mathcal{E}}{\hbar} = \frac{1}{\hbar} \left(m_ec^2 + \frac{\mathcal{E}_b}{2} \right) \qquad (S.2.12.3)$$

The velocity of the photon is, of course, $u = \pm c$.

b) To find the velocity and the frequency of the photon in the lab frame we can use the Lorentz transformations for momentum and energy for the photon that is registered by the observer (with momentum p_- from (S.2.12.2)):

$$p' = \gamma \left(p_- + \frac{v}{c^2} \mathcal{E} \right)$$

$$\mathcal{E}' = \gamma(\mathcal{E} + vp_-)$$

where $\mathcal{E}, p$ are the energy and momentum of the photon in the rest frame, and $\mathcal{E}', p'$ are the energy and momentum of the photon in the observer's frame.

$$p' = \gamma \left[-\left(m_e c + \frac{\mathcal{E}_b}{2c} \right) + \frac{v}{c^2} \left(m_e c^2 + \frac{\mathcal{E}_b}{2} \right) \right] = \frac{\gamma}{c} \left(m_e c^2 + \frac{\mathcal{E}_b}{2} \right) (\beta - 1)$$

$$\mathcal{E}' = \gamma \left[m_e c^2 + \frac{\mathcal{E}_b}{2} - v \left(m_e c + \frac{\mathcal{E}_b}{2c} \right) \right] = \gamma \left(m_e c^2 + \frac{\mathcal{E}_b}{2} \right) (1 - \beta)$$

so the velocity of the photon in the observer's frame is $u' = p' c^2 / \mathcal{E}' = -c$, which of course is what one would expect. The frequency of the photon measured by the observer is

$$\omega' = \frac{\mathcal{E}'}{\hbar} = \frac{\gamma}{\hbar} \left(m_e c^2 + \frac{\mathcal{E}_b}{2c} \right) (1 - \beta) = \gamma \omega (1 - \beta)$$

where we substitute ω from (S.2.12.3) of (a). So the original frequency ω is redshifted to

$$\omega' = \omega \frac{1 - \beta}{\sqrt{1 - \beta^2}} = \omega \sqrt{\frac{1 - \beta}{1 + \beta}}$$

2.13 Transverse Relativistic Doppler Effect (Moscow Phys-Tech)

Intuitively it is clear that the effect is due to a time dilation $t' = \gamma t$ so $\omega' = \omega / \gamma$ ($\omega \propto 1/t$). More formally, we can use the energy and momentum transformation as in Problem 2.12, but it is more convenient to introduce a 4-vector k^μ (**k** is a wave vector)

$$k^\mu = \left(\frac{\omega}{c}, \mathbf{k} \right)$$

and consider its transformation from the rest frame K to the observer's frame K'

$$k^0 = \gamma(k'^0 - \beta k'^1) \tag{S.2.13.1}$$

where $k'^0 = \omega'/c$, $k'^1 = k' \cos \alpha = 0$ ($\alpha = \pi/2$). Substituting this into

(S.2.13.1), we find

$$k^0 = \frac{\omega}{c} = \gamma \frac{\omega'}{c}$$

so

$$\omega' = \frac{\omega}{\gamma} = \omega\sqrt{1 - \beta^2}$$

Note that the transverse Doppler effect gives only a second order correction to the frequency

$$\omega' \approx \omega - \frac{1}{2}\omega\beta^2$$

where $\beta \ll 1$, whereas the longitudinal Doppler effect yields a first order correction (see Problem 2.12).

2.14 Particle Creation (MIT)

In this problem and in most of the problems involving particle reactions, we will use units where $c = 1$. Compose the invariant quantity for the system $p^\mu p_\mu$. Before the reaction in the rest frame of the proton

$$p^\mu = (m_p, 0) + (\mathcal{E}_\gamma, \mathbf{p}_\gamma) = (m_p + \mathcal{E}_\gamma, \mathbf{p}_\gamma) \qquad (S.2.14.1)$$

$$p^\mu p_\mu = m_p^2 + \mathcal{E}_\gamma^2 + 2m_p\mathcal{E}_\gamma - \mathcal{E}_\gamma^2 = m_p^2 + 2m_p\mathcal{E}_\gamma \qquad (S.2.14.2)$$

where we used $p_\gamma = \mathcal{E}_\gamma$. After the reaction, in the center of mass frame, the lowest energy products will take away no momentum, so

$$p^\mu p_\mu = (m_p + m_\pi)^2 = m_p^2 + m_\pi^2 + 2m_p m_\pi \qquad (S.2.14.3)$$

Setting (S.2.14.2) equal to (S.2.14.3), we find $\mathcal{E}_\gamma = \mathcal{E}_{\min}$.

$$2m_p\mathcal{E}_{\min} = m_\pi^2 + 2m_p m_\pi \qquad (S.2.14.4)$$

So the threshold for this reaction

$$\mathcal{E}_{\min} = \frac{m_\pi(m_\pi + 2m_p)}{2m_p} = m_\pi\left(1 + \frac{m_\pi}{2m_p}\right) \qquad (S.2.14.5)$$

2.15 Electron–Electron Collision (Stony Brook)

a) The total energy in the laboratory frame

$$\mathcal{E} = \mathcal{E}_1 + \mathcal{E}_2 = \mathcal{E}_1 + mc^2 = 1.40 \text{ MeV} + 0.51 \text{ MeV} = 1.91 \text{ MeV}$$

where $\mathcal{E}_1$ is the energy of the moving electron and $\mathcal{E}_2 = mc^2$ is the energy of the electron at rest. The total momentum is simply the momentum p_1 of the moving electron

$$P = p_1 = \sqrt{\mathcal{E}_1^2/c^2 - m^2c^2} = \sqrt{(1.40)^2 - (0.51)^2}\ \text{MeV}/c = 1.30\ \text{MeV}/c$$

b) The velocity of the center of mass in the lab frame is defined as

$$\mathbf{v} = \frac{\mathbf{P}}{\mathcal{E}/c^2} = \frac{c^2\mathbf{P}}{\mathcal{E}}$$

where $\mathbf{P}$ and $\mathcal{E}$ are the total momentum and energy in the lab frame, respectively.

$$v = |\mathbf{v}| = \frac{1.30}{1.91} = 0.68c$$

c) The total energy in the CMF $\mathcal{E}'$ ($P' = 0$) is given by

$$\mathcal{E}'^2 - P'^2c^2 = \mathcal{E}^2 - P^2c^2$$

so

$$\mathcal{E}' = (\mathcal{E}^2 - P^2c^2)^{1/2} = 1.40\ \text{MeV}$$

We may also calculate the total energy in the CMF by transforming the energy of the target from the laboratory frame to the CMF:

$$\mathcal{E}_2' = \gamma\mathcal{E}_2 + \gamma\beta cp_2 = \gamma m$$

Since the target and the projectile have the same energy in the CMF, the total energy is given by

$$\mathcal{E}' = 2\mathcal{E}_2' = 2\gamma m = \frac{2 \cdot 0.51}{\sqrt{1 - (0.68)^2}} = 1.40\ \text{MeV}$$

d) In the center of mass frame the total momentum $\mathbf{P} = 0$, therefore, $\mathbf{p}_1 = -\mathbf{p}_2$ and the angle of the scatter $\alpha = 135°$. The momentum can be found by considering the 4-momentum of the two particles. We will use another method here. Before the collision the projectile had a momentum p_1 in the lab frame; in the CMF it had a momentum

$$p_1 = \gamma\left(p_1 - \frac{v}{c^2}\mathcal{E}_1\right)$$

$\mathbf{p}_1$ and $\mathbf{p}_2'$ are both along the $\hat{\mathbf{x}}$ axis

$$p_1' = \frac{1}{\sqrt{1 - (0.68)^2}}(1.30 - 0.68 \cdot 1.40) = 0.47\ \text{MeV}/c$$

As a result of the collision, the momentum of the projectile will only rotate, and the momentum of the target

$$|\mathbf{p}_2'| = |\mathbf{p}_1'| = 0.47 \text{ MeV}/c$$

The energy of the target $\mathcal{E} = \sqrt{p_1'^2 + m^2 c^4} = 0.70$ MeV, or from the same transformation as that of the projectile

$$\mathcal{E}_1' = \gamma(\mathcal{E}_1 - v p_1) = \frac{1}{\sqrt{1 - (0.68)^2}} (1.40 - 0.68 \cdot 1.30) = 0.70 \text{ MeV}$$

which is, of course, the same as above, where we used the fact that the energy of the projectile is equal to the target energy, since the masses of the two particles are the same.

e) In the $\hat{\mathbf{y}}$ direction perpendicular to the direction of the incoming particle

$$p_y = p_y' = p_2 \sin \chi = \frac{0.47 \text{ MeV}/c}{\sqrt{2}} = 0.34 \text{ MeV}/c$$

To find the $\hat{\mathbf{x}}$ component in the lab frame we can again use the Lorentz transformation of momentum

$$p_x = \gamma \left(p_x' + \frac{v}{c^2} \mathcal{E} \right) = \frac{1}{\sqrt{1 - (0.68)^2}} (0.34 + 0.68 \cdot 0.70) = 1.11 \text{ MeV}/c$$

2.16 Inverse Compton Scattering (MIT, Maryland)

Write the energy and momentum conservation laws in 4-notation, where we use units in which $c = 1$:

$$p_\gamma^\mu + p_e^\mu = \tilde{p}_\gamma^\mu + \tilde{p}_e^\mu \tag{S.2.16.1}$$

where p_γ, p_e, and $\tilde{p}_\gamma$, $\tilde{p}_e$ are the 4-momenta of the photon and electron before and after the collision, respectively. After subtracting $\tilde{p}_\gamma^\mu$ from both sides of (S.2.16.1) and squaring, we obtain

$$p_\gamma^\mu p_{e\mu} - p_\gamma^\mu \tilde{p}_{\gamma\mu} - p_e^\mu \tilde{p}_{\gamma\mu} = 0 \tag{S.2.16.2}$$

Here we used $p_e^\mu p_{e\mu} = \tilde{p}_e^\mu \tilde{p}_{e\mu} = m_e^2$ and $p_\gamma^\mu p_{\gamma\mu} = 0$. We may rewrite (S.2.16.2) in the form

$$\mathcal{E}_\gamma \mathcal{E}_e - \mathbf{p}_\gamma \cdot \mathbf{p}_e - \mathcal{E}_\gamma \tilde{\mathcal{E}}_\gamma + \mathbf{p}_\gamma \cdot \tilde{\mathbf{p}}_\gamma - \mathcal{E}_e \tilde{\mathcal{E}}_\gamma + \mathbf{p}_e \cdot \tilde{\mathbf{p}}_\gamma = 0 \tag{S.2.16.3}$$

Introducing the angle θ between the direction of the incident and scattered photons, we have

$$\mathcal{E}_\gamma p_e + \mathcal{E}_\gamma \mathcal{E}_e - \mathcal{E}_\gamma \tilde{\mathcal{E}}_\gamma + \mathcal{E}_\gamma \tilde{\mathcal{E}}_\gamma \cos\theta - \mathcal{E}_e \tilde{\mathcal{E}}_\gamma - p_e \tilde{\mathcal{E}}_\gamma \cos\theta = 0 \qquad \text{(S.2.16.4)}$$

From (S.2.16.4)

$$\tilde{\mathcal{E}}_\gamma = \frac{\mathcal{E}_\gamma (p_e + \mathcal{E}_e)}{\mathcal{E}_\gamma (1 - \cos\theta) + \mathcal{E}_e + p_c \cos\theta} \qquad \text{(S.2.16.5)}$$

The maximum energy $\tilde{\mathcal{E}}_{\gamma\,\text{max}}$ in (S.2.16.5) corresponds to the minimum of the denominator which yields $\cos\theta = -1$, $\theta = \pi$ (backscattering). Therefore

$$\tilde{\mathcal{E}}_{\gamma\,\text{max}} = \frac{\mathcal{E}_\gamma (p_e + \mathcal{E}_e)}{2\mathcal{E}_\gamma + \mathcal{E}_e - p_e} \qquad \text{(S.2.16.6)}$$

Using the fact that $m_e \ll \mathcal{E}_e$ ($p_e \approx \mathcal{E}_e$) and expanding the denominator, we can rewrite (S.2.16.6) in the form

$$\tilde{\mathcal{E}}_{\gamma\,\text{max}} \approx \frac{2\mathcal{E}_\gamma}{2\mathcal{E}_\gamma/\mathcal{E}_e + (1 - p_e/\mathcal{E}_e)} \approx \frac{\mathcal{E}_\gamma}{\mathcal{E}_\gamma/\mathcal{E}_e + m_e^2/4\mathcal{E}_e^2} \qquad \text{(S.2.16.7)}$$

We may now substitute values into (S.2.16.7): $\mathcal{E}_\gamma = hc/\lambda = 2.42$ eV, $m_e = 0.51$ MeV, and $\mathcal{E}_e \approx p_e = 27$ GeV, so

$$\tilde{\mathcal{E}}_{\gamma\,\text{max}} = \frac{2.42 \text{ eV}}{\dfrac{2.42}{27 \cdot 10^9} + \left(\dfrac{0.51}{54 \cdot 10^3}\right)^2} = \frac{2.42 \text{ eV}}{9.0 \cdot 10^{-11} + 8.9 \cdot 10^{-11}} = 13.5 \text{ GeV}$$

2.17   Proton–Proton Collision (MIT)

Since the energies of the protons after the collision are equal, they will rebound at the same angle $\theta/2$ relative to the initial momentum of the proton (see Figure S.2.17). Again we use $c = 1$. In these units before the

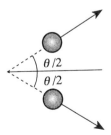

Figure S.2.17

collision

$$p = m\beta\gamma$$

$$E = m\gamma$$

Using momentum conservation, we have

$$m\beta\gamma = 2m\tilde{\beta}\tilde{\gamma}\cos\left(\frac{\theta}{2}\right) \qquad (S.2.17.1)$$

or $\qquad\qquad \beta\gamma = 2\tilde{\beta}\tilde{\gamma}\cos\dfrac{\theta}{2} \qquad\qquad\qquad (S.2.17.2)$

where $\tilde{\beta}$ and $\tilde{\gamma}$ stand for β and γ after the collision. Energy conservation yields

$$m + m\gamma = 2m\tilde{\gamma} \qquad (S.2.17.3)$$

or

$$\gamma + 1 = 2\tilde{\gamma} \qquad (S.2.17.4)$$

Now,

$$\beta\gamma = \sqrt{\gamma^2 - 1} \qquad (S.2.17.5)$$

So, we obtain by using (S.2.17.4)

$$\tilde{\beta}\tilde{\gamma} = \sqrt{\tilde{\gamma}^2 - 1} = \sqrt{(\gamma+1)^2/4 - 1} \qquad (S.2.17.6)$$

Substituting (S.2.17.6) into (S.2.17.2) and using (S.2.17.5) gives

$$\sqrt{\gamma^2 - 1} = 2\sqrt{(\gamma+1)^2/4 - 1}\,\cos\frac{\theta}{2}$$

$$\gamma^2 - 1 = (\gamma^2 + 2\gamma - 3)\cos^2\frac{\theta}{2} = (\gamma-1)(\gamma+3)\cos^2\frac{\theta}{2}$$

$$\cos\theta = 2\cos^2\frac{\theta}{2} - 1 = \frac{\gamma-1}{\gamma+3}$$

For $\gamma \approx 1$, i.e., in the classical limit of low velocity, $\cos\theta = 0$, and we obtain the familiar result that the angle between billiard balls rebounding with equal energy is 90°. If $\gamma \gg 1$ (extremely relativistic case), then $\cos\theta \approx 1$ and $\theta \to 0$.

2.18 Pion Creation and Neutron Decay (Stony Brook)

a) The threshold energy for the creation of a neutron n and a pion π^+ in the center of mass frame is simply the sum of their masses (in units where

$c = 1$). The 4-momentum of the neutron/pion system in the center of mass frame is

$$P'^{\mu}_{n\pi} = (m_n + m_{\pi^+}, 0) \tag{S.2.18.1}$$

We must also calculate the energy of the incident particles in the center of mass frame. However, we may use the relativistic invariance of a product of 4-vectors (which can be written for a complex system as well as for a single particle). The 4-momentum in the lab (stationary proton) frame is

$$P^{\mu}_{\gamma} = (\mathcal{E}_{\gamma}, \mathbf{p}_{\gamma}) \qquad\qquad P^{\mu}_{p} = (m_p, 0)$$

$$P^{\mu}_{p\gamma} = (\mathcal{E}_{\gamma} + m_p, \mathbf{p}_{\gamma}) \tag{S.2.18.2}$$

Taking the squares of (S.2.18.1) and (S.2.18.2) and equating them, we obtain

$$(\mathcal{E}_{\gamma} + m_p)^2 - \mathbf{p}_{\gamma}^2 = (m_n + m_{\pi^+})^2 \tag{S.2.18.3}$$

Note that in these units $|\mathbf{p}_{\gamma}| = \mathcal{E}_{\gamma}$. Therefore, we have for the threshold energy

$$\mathcal{E}_{\gamma} = \frac{(m_n + m_{\pi^+})^2 - m_p^2}{2m_p} \tag{S.2.18.4}$$

Substituting the values given in the problem

$$\mathcal{E}_{\gamma} = \frac{(939.57 + 139.57)^2 - (938.27)^2}{2 \cdot (938.27)} \text{ MeV} = 151.4 \text{ MeV}$$

b) Consider the neutron decay following the scheme

$$n \rightarrow p\bar{\nu}_e + e^-$$

where we consider the proton and neutrino as one complex particle. From energy and momentum conservation, we find

$$0 = p_{p\bar{\nu}_e} + p_{e^-} \tag{S.2.18.5}$$

$$m_n = \mathcal{E}_{p\bar{\nu}_e} + \mathcal{E}_{e^-} \tag{S.2.18.6}$$

$$\mathcal{E}_{p\bar{\nu}_e} = \sqrt{p_{p\bar{\nu}_e}^2 + m_{p\bar{\nu}_e}^2} \tag{S.2.18.7}$$

$$\mathcal{E}_{e^-} = \sqrt{p_{e^-}^2 + m_{e^-}^2} \tag{S.2.18.8}$$

Substituting (S.2.18.6) into (S.2.18.8), we have

$$p_{e^-}^2 = (m_n - \mathcal{E}_{p\bar{\nu}_e})^2 - m_{e^-}^2$$

(S.2.18.9)

$$p_{e^-} = \sqrt{(m_n - \mathcal{E}_{p\bar{\nu}_e})^2 - m_{e^-}^2}$$

We are looking for the maximum possible momentum p_{e^-} of the electron. It depends only on one variable, $\mathcal{E}_{p\bar{\nu}_e}$, where $\mathcal{E}_{p\bar{\nu}_e} < m_n$, so the maximum p_{e^-} corresponds to a minimum $\mathcal{E}_{p\bar{\nu}_e}$. Using (S.2.18.5) and (S.2.18.7), we obtain

$$\mathcal{E}_{p\bar{\nu}_e}^2 = m_{p\bar{\nu}_e}^2 + p_{p\bar{\nu}_e}^2 = m_{p\bar{\nu}_e}^2 + p_{e^-}^2$$

(S.2.18.10)

Substituting $p_{e^-}^2$ from (S.2.18.9) into (S.2.18.10) yields

$$0 = m_n^2 - 2m_n \mathcal{E}_{p\bar{\nu}_e} + m_{p\bar{\nu}_e}^2 - m_{e^-}^2$$

(S.2.18.11)

$$\mathcal{E}_{p\bar{\nu}_e} = \frac{m_n^2 + m_{p\bar{\nu}_e}^2 - m_{e^-}^2}{2m_n}$$

From this equation, the minimum of $\mathcal{E}_{p\bar{\nu}_e}$ corresponds to a minimum of $m_{p\bar{\nu}_e}^2$. Now,

$$m_{p\bar{\nu}_e}^2 = (\mathcal{E}_p + \mathcal{E}_\nu)^2 - (\mathbf{p}_p + \mathbf{p}_\nu)^2$$

$$= \mathcal{E}_p^2 + 2\mathcal{E}_p\mathcal{E}_\nu - p_p^2 - 2p_p p_\nu \cos\theta$$

$$= m_p^2 + 2\mathcal{E}_\nu(\mathcal{E}_p - p_p \cos\theta)$$

(S.2.18.12)

The minimum of (S.2.18.12) corresponds to $\mathcal{E}_\nu = 0$, so $m_{p\bar{\nu}_e} = m_p$. Therefore

$$|\mathbf{P}_e|_{max} = \sqrt{\mathcal{E}_{p\bar{\nu}_e}^2 - m_{p\bar{\nu}_e}^2} = \sqrt{\left(\frac{m_n^2 + m_p^2 - m_{e^-}^2}{2m_n}\right)^2 - m_p^2}$$

(S.2.18.13)

$$= 1.19 \text{ MeV}$$

where we have substituted the masses in energy units. On the other hand

$$p_{e\,max} = m_e \beta \gamma = \frac{m_e \beta}{\sqrt{1 - \beta^2}}$$

(S.2.18.14)

From (S.2.18.14), we have

$$\beta = \frac{p_{e\,\max}/m_e}{\sqrt{1+(p_{e\,\max}/m_e)^2}} = \frac{1.19/0.51}{\sqrt{1+(1.19/0.51)^2}} = 0.92$$

Therefore in regular units

$$v_{e\,\max} = \beta c = 0.92c$$

2.19   Elastic Collision and Rotation Angle (MIT)

a) Express conservation of energy as

$$\mathcal{E}_{10} + \mathcal{E}_{20} = \tilde{\mathcal{E}}_{10} + \tilde{\mathcal{E}}_{20} \tag{S.2.19.1}$$

Conservation of momentum gives

$$\tilde{\mathbf{p}}_{10} = -\tilde{\mathbf{p}}_{20} \equiv \tilde{\mathbf{p}}_0 \tag{S.2.19.2}$$

From the information given and (S.2.19.2)

$$\mathcal{E}_{10}^2 - p_0^2 = m_1^2 \qquad \mathcal{E}_{20}^2 - p_0^2 = m_2^2$$

$$\tilde{\mathcal{E}}_{10}^2 - \tilde{p}_0^2 = m_1^2 \qquad \tilde{\mathcal{E}}_{20}^2 - \tilde{p}_0^2 = m_2^2 \tag{S.2.19.3}$$

These equations may be rearranged to yield

$$\mathcal{E}_{10}^2 - \mathcal{E}_{20}^2 = \tilde{\mathcal{E}}_{10}^2 - \tilde{\mathcal{E}}_{20}^2$$

$$(\mathcal{E}_{10} - \mathcal{E}_{20})(\mathcal{E}_{10} + \mathcal{E}_{20}) = (\tilde{\mathcal{E}}_{10} - \tilde{\mathcal{E}}_{20})(\tilde{\mathcal{E}}_{10} + \tilde{\mathcal{E}}_{20}) \tag{S.2.19.4}$$

Substituting (S.2.19.1) into (S.2.19.4) yields

$$\mathcal{E}_{10} - \mathcal{E}_{20} = \tilde{\mathcal{E}}_{10} - \tilde{\mathcal{E}}_{20} \tag{S.2.19.5}$$

Adding or subtracting (S.2.19.1) from (S.2.19.5) results in

$$\mathcal{E}_{10} = \tilde{\mathcal{E}}_{10} \qquad \mathcal{E}_{20} = \tilde{\mathcal{E}}_{20} \tag{S.2.19.6}$$

and by (S.2.19.3),

$$|\mathbf{p}_0| = |\tilde{\mathbf{p}}_0| \tag{S.2.19.7}$$

b) Inspection of (P.2.19.3) seems to indicate the way to proceed, since $\tilde{p}_2$ does not appear. Subtract $\tilde{p}_{1\mu}$ from both sides of (P.2.19.2) and square:

$$(p_1 + p_2 - \tilde{p}_1) \cdot (p_1 + p_2 - \tilde{p}_1) = \tilde{p}_2 \cdot \tilde{p}_2$$

$$= m_1^2 + m_2^2 + m_1^2 + 2p_1 \cdot p_2 - 2p_1 \cdot \tilde{p}_1 - 2\tilde{p}_1 \cdot p_2 = m_2^2$$

$$p_2 \cdot p_1 - p_1 \cdot \tilde{p}_1 - \tilde{p}_1 \cdot p_2 + m_1^2 = 0 \qquad \text{(S.2.19.8)}$$

c) The instructions in (c) exploit the invariance of a product of 4-vectors under a Lorentz transformation, here from the laboratory frame to the CMF. In the laboratory frame

$$p_1 = (\mathcal{E}_1, \mathbf{p}_1) \qquad \qquad \tilde{p}_1 = (\tilde{\mathcal{E}}_1, \tilde{\mathbf{p}}_1)$$

$$p_2 = (m_2, 0)$$

So we have

$$p_1 \cdot p_2 = \mathcal{E}_1 m_2$$

$$p_2 \cdot \tilde{p}_1 = \tilde{\mathcal{E}}_1 m_2$$

In the CMF

$$p_1 = (\mathcal{E}_{10}, \mathbf{p}_0) \qquad \tilde{p}_1 = (\tilde{\mathcal{E}}_{10}, \tilde{\mathbf{p}}_0)$$

where

$$|\mathbf{p}_0| = |\tilde{\mathbf{p}}_0|$$

$$p_2 = (\mathcal{E}_{20}, -\mathbf{p}_0)$$

$$p_1 \cdot \tilde{p}_1 = \mathcal{E}_{10}^2 - p_0^2 \cos \chi$$

Substitution into (S.2.19.8) yields

$$\mathcal{E}_1 m_2 - \mathcal{E}_{10}^2 + p_0^2 \cos \chi - \tilde{\mathcal{E}}_1 m_2 + m_1^2 = 0$$

Now, $\mathcal{E}_{10}^2 = p_0^2 + m_1^2$, so

$$(\tilde{\mathcal{E}}_1 - \mathcal{E}_1) m_2 = -p_0^2 (1 - \cos \chi)$$

$$\tilde{\mathcal{E}}_1 - \mathcal{E}_1 = -\frac{p_0^2}{m_2} (1 - \cos \chi) \qquad \text{(S.2.19.9)}$$

d) From (c), $p_1 \cdot p_2$ in the laboratory frame equals $\mathcal{E}_1 m_2$, and $p_1 \cdot p_2$ in the CMF equals $\mathcal{E}_{10}\mathcal{E}_{20} + p_0^2$, so

$$\mathcal{E}_1 m_2 = \mathcal{E}_{10}\mathcal{E}_{20} + p_0^2 \qquad (S.2.19.10)$$

$$\mathcal{E}_{10} = \sqrt{p_0^2 + m_1^2}$$

$$\mathcal{E}_{20} = \sqrt{p_0^2 + m_2^2}$$

Rearrange and square (S.2.19.10):

$$(\mathcal{E}_1 m_2 - p_0^2)^2 = (p_0^2 + m_1^2)(p_0^2 + m_2^2) \qquad (S.2.19.11)$$

yielding

$$p_0^2(m_1^2 + m_2^2 + 2\mathcal{E}_1 m_2) = m_2^2(\mathcal{E}_1^2 - m_1^2)$$

$$p_0^2 = m_2^2 \frac{\mathcal{E}_1^2 - m_1^2}{m_1^2 + m_2^2 + 2\mathcal{E}_1 m_2}$$

$$p_0 = m_2 \sqrt{\frac{\mathcal{E}_1^2 - m_1^2}{m_1^2 + m_2^2 + 2\mathcal{E}_1 m_2}} \qquad (S.2.19.12)$$

If the incident mass has no kinetic energy, in the CMF it should have no momentum, as seen in (S.2.19.12). If $m_2/m_1 \to 0$, then the frame of particle 1 is the CMF, so p_0 again should equal 0.

e) From (S.2.19.9) and (S.2.19.12) we have

$$\tilde{\mathcal{E}}_1 = \mathcal{E}_1 - m_2 \frac{(\mathcal{E}_1^2 - m_1^2)(1 - \cos\chi)}{m_1^2 + m_2^2 + 2\mathcal{E}_1 m_2}$$

Conservation of energy gives

$$\mathcal{E}_1 + \mathcal{E}_2 = \tilde{\mathcal{E}}_1 + \tilde{\mathcal{E}}_2$$

So

$$\tilde{\mathcal{E}}_2 = \mathcal{E}_1 + \mathcal{E}_2 - \tilde{\mathcal{E}}_1$$

$$\tilde{\mathcal{E}}_2 = m_2 + m_2 \frac{(\mathcal{E}_1^2 - m_1^2)(1 - \cos\chi)}{m_1^2 + m_2^2 + 2\mathcal{E}_1 m_2}$$

f) We maximize $\tilde{\mathcal{E}}_2(\chi)$ by letting $\chi = \pi$. So

$$\tilde{\mathcal{E}}_2 = m_2 \frac{(\mathcal{E}_1 + m_2)^2 + (\mathcal{E}_1^2 - m_1^2)}{m_1^2 + m_2^2 + 2\mathcal{E}_1 m_2}$$

$$\tilde{\mathcal{E}}_1 = \mathcal{E}_1 - m_2 \frac{2(\mathcal{E}_1^2 - m_1^2)}{m_1^2 + m_2^2 + 2\mathcal{E}_1 m_2}$$

$$\frac{\tilde{\mathcal{E}}_1 - m_1}{\mathcal{E}_1 - m_1} = \frac{(\mathcal{E}_1 - m_1)(m_1^2 + m_2^2 + 2\mathcal{E}_1 m_2) - 2m_2(\mathcal{E}_1^2 - m_1^2)}{(\mathcal{E}_1 - m_1)(m_1^2 + m_2^2 + 2\mathcal{E}_1 m_2)}$$

$$= \frac{m_1^2 + m_2^2 + 2\mathcal{E}_1 m_2 - 2m_2(\mathcal{E}_1 + m_1)}{m_1^2 + m_2^2 + 2\mathcal{E}_1 m_2}$$

$$= \frac{(m_1 - m_2)^2}{m_1^2 + m_2^2 + 2\mathcal{E}_1 m_2}$$

For $m_1 = m_2$, the final kinetic energy of the incident particle equals 0.

3

Electrodynamics

3.1 Charge Distribution (Wisconsin-Madison)

We can use Gauss's theorem in its integral form:

$$\int_{S_0} \mathbf{E} \cdot d\mathbf{S} = 4\pi Q$$

where Q is the net charge inside the surface S_0.

$$Q = \frac{1}{4\pi} \int_{S_0} \mathbf{E} \cdot d\mathbf{S} = \frac{c}{4\pi} \int_{S_0} \left(1 - e^{-\alpha r}\right) \frac{\hat{\mathbf{r}}}{r^2} \cdot d\mathbf{S}$$

$$= \frac{c}{4\pi} \left(1 - e^{-\alpha(1/\alpha)}\right) \alpha^2 \frac{4\pi}{\alpha^2} = c\left(1 - e^{-1}\right)$$

3.2 Electrostatic Forces and Scaling (Moscow Phys-Tech)

a) The charged sphere will polarize the neutral one, which acquires a dipole moment p proportional to the electric field created by the charged sphere

$$p \propto E \propto \frac{q}{R^2}$$

The force between the dipole and the charged sphere is given by the product of the dipole moment and the gradient of the electric field at the dipole:

$$F \propto \frac{pq}{R^3} \propto \frac{q^2}{R^5}$$

201

The condition that $F = F'$ after increasing the distance by a factor of two gives for the new charge q'

$$\frac{q^2}{R^5} = \frac{q'^2}{R'^5}$$

So

$$\left(\frac{q'}{q}\right)^2 = \left(\frac{R'}{R}\right)^5 = 32$$

$$q' = 4\sqrt{2}\, q$$

b) The charge Q will be distributed uniformly along the wire. So the Coulomb force between some point on the ring and the rest of the ring will be proportional to the square of Q and inversely proportional to the square of the diameter D of the ring

$$F_Q \propto \frac{Q^2}{D^2} \qquad\qquad (S.3.2.1)$$

When the ring breaks, the elastic force F_e attempting to maintain the integrity of the ring is given by

$$F_e = \sigma S \qquad\qquad (S.3.2.2)$$

where σ is the ultimate strength, which depends only on the material of the wire, and S is the cross section of the wire. At the point when the ring parts, $F_Q = F_e$, so equating (S.3.2.1) and (S.3.2.2), we obtain

$$Q^2 \propto \sigma S D^2 \propto \sigma d^2 D^2$$

Scaling up the linear dimensions by a factor of two gives

$$Q' \propto d'D' = 4dD$$

Therefore

$$Q' = 4Q$$

3.3 Dipole Energy (MIT, Moscow Phys-Tech)

a) The dipole is attracted to the plane, as seen from the position of the image charges (see Figure S.3.3).

b) The field at a point $\mathbf{r}$ due to a dipole at the origin is given by

$$\mathbf{E}(\mathbf{r}) = \frac{3\mathbf{r}(\mathbf{r} \cdot \mathbf{p}) - \mathbf{p}r^2}{r^5} \qquad\qquad (S.3.3.1)$$

The potential energy U of the dipole in the field of another dipole is given

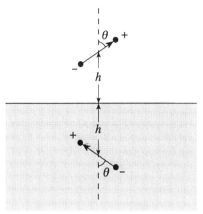

F i g u r e   S.3.3

by $-\mathbf{p} \cdot \mathbf{E}$. Therefore, for two dipoles $\mathbf{p}_1$ and $\mathbf{p}_2$

$$U = \frac{-3(\mathbf{r} \cdot \mathbf{p}_2)(\mathbf{r} \cdot \mathbf{p}_1) + (\mathbf{p}_1 \cdot \mathbf{p}_2)\, r^2}{r^5} \qquad (S.3.3.2)$$

where $\mathbf{r}$ is the vector from dipole 1 to dipole 2. Extra care must be exercised here since this is an image problem and not one where a single dipole remains fixed and the other is brought to infinity. The force F on the dipole equals $-\partial U/\partial r$, and we want to integrate $F\, dx$ from $x = h$ to ∞, where $x = r/2$. The work done

$$W = \int\limits_{x=h}^{\infty} -\left(\frac{\partial U}{\partial r}\right) dx = -\int\limits_{x=h}^{\infty} \left(\frac{\partial U}{\partial r}\right) dr \frac{\partial x}{\partial r} = -\frac{1}{2} \int\limits_{r=2h}^{\infty} \frac{\partial U}{\partial r}\, dr$$

$$= -\frac{1}{2}\, U(r) \Big|_{2h}^{\infty} = \frac{1}{2} \frac{3(2h)^2 p^2 \cos^2 \theta - (2h)^2 p^2 \cos 2\theta}{(2h)^5} \qquad (S.3.3.3)$$

$$= \frac{1}{2} \frac{(1 + \cos^2 \theta)p^2}{(2h)^3}$$

The work necessary to move the dipole to infinity from a real fixed dipole is twice that found in (S.3.3.3).

3.4   Charged Conducting Sphere in Constant Electric Field (Stony Brook, MIT)

Look for a solution of the form

$$\phi = \phi_0 + \phi_1 + \frac{Q}{r}$$

where $\phi_0 = -\mathbf{E}_0 \cdot \mathbf{r}$ is the potential due to the external field and ϕ_1 is the change in the potential due to the presence of the sphere. The constant vector $\mathbf{E}_0$ defines a preferred direction, and therefore the potential ϕ_1 may depend only on this vector. Then, the only solution of Laplace's equation which goes to zero at infinity is a dipole potential (see, for instance, Landau and Lifshitz, *Electrodynamics of Continuous Media*, §3)

$$\phi_1 = -A\mathbf{E}_0 \cdot \nabla \frac{1}{r} = A\frac{\mathbf{E}_0 \cdot \mathbf{r}}{r^3} \qquad (\text{S.3.4.1})$$

where A is some constant (alternatively, we may write the solution in terms of Legendre polynomials and obtain the same answer from the boundary conditions). So

$$\phi = -\mathbf{E}_0 \cdot \mathbf{r} + A\frac{\mathbf{E}_0 \cdot \mathbf{r}}{r^3} + \frac{Q}{r} \qquad (\text{S.3.4.2})$$

On the surface of the sphere, ϕ is constant:

$$\phi = -E_0 a \cos\theta + A\frac{E_0 a \cos\theta}{a^3} + \frac{Q}{a} = \text{const} \qquad (\text{S.3.4.3})$$

where θ is the angle between $\mathbf{E}_0$ and $\mathbf{r}$ (see Figure S.3.4). From (S.3.4.3), we find that $A = a^3$, and finally

$$\phi = -E_0 r \cos\theta \left(1 - \frac{a^3}{r^3}\right) + \frac{Q}{r} \qquad (\text{S.3.4.4})$$

The surface charge density

$$\sigma = \frac{1}{4\pi}E_{\text{n}} = -\frac{1}{4\pi}\frac{\partial\phi}{\partial n} = -\frac{1}{4\pi}\frac{\partial\phi}{\partial r}\bigg|_{r=a} = \frac{3}{4\pi}E_0\cos\theta + \frac{Q}{4\pi a^2}$$

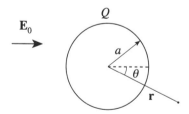

Figure   S.3.4

3.5   Charge and Conducting Sphere I (MIT)

a) First replace the sphere by an image charge that will create zero potential on the surface of the sphere. We know that is possible to do so with only

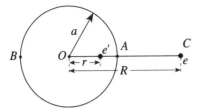

Figure   S.3.5a

one image charge e' since we can always find a spherical surface of zero potential for two charges (see Problem 3.18). In general, we must consider the potential at arbitrary points on the surface. Consider, for simplicity, two points A and B on opposite sides of a diameter (see Figure S.3.5a). The potentials at points A and B due to the two charges e and e', are, respectively,

$$\phi_A = \frac{e}{R-a} + \frac{e'}{a-r} = 0 \qquad (S.3.5.1)$$

and

$$\phi_B = \frac{e}{R+a} + \frac{e'}{r+a} = 0 \qquad (S.3.5.2)$$

or

$$e(a-r) + e'(R-a) = 0 \qquad (S.3.5.3)$$

$$e(r+a) + e'(R+a) = 0 \qquad (S.3.5.4)$$

From (S.3.5.3) and (S.3.5.4), we can find e' and r

$$e' = -\frac{ea}{R}$$
$$r = \frac{a^2}{R}$$

For a neutral sphere, the total charge is constant ($Q = 0$), so we have to add yet another charge $e'' = -e'$ and at the same time keep the potential constant on the surface of the sphere (see Figure S.3.5b). Obviously, we

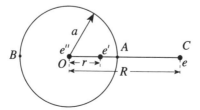

Figure   S.3.5b

must put this charge at the center of the sphere. The potential on the surface of the sphere is therefore

$$\phi_0 = \frac{e''}{a} = -\frac{e'}{a} = \frac{ea}{Ra} = \frac{e}{R}$$

(since the potential due to the other two charges is zero).

b) The force can be found from the interaction between the charge e and the two image charges e' and e''. The force is attractive and directed along the radius vector to e.

$$F = \frac{ee'}{(R-r)^2} + \frac{ee''}{R^2} = -\frac{e^2 a}{R(R-r)^2} + \frac{e^2 a}{R^3} = \frac{e^2 a}{R}\left(-\frac{1}{(R-a^2/R)^2} + \frac{1}{R^2}\right)$$

$$= -\frac{e^2 a^3 \left(2R^2 - a^2\right)}{R^3 (R^2 - a^2)^2}$$

3.6 Charge and Conducting Sphere II (Boston)

In Problem 3.5, we found the expression for the force acting on the charge in the presence of an insulated conducting sphere. For the grounded sphere, the potential on the surface of the sphere is zero, and therefore there is only one image charge

$$e' = -\frac{ea}{R}$$

at a distance

$$r = \frac{a^2}{R}$$

from the center of the sphere. The force on e can be computed as the force between it and the image charge e':

$$F = \frac{ee'}{(R-r)^2} = -\frac{e^2 a}{R(R - a^2/R)^2} = -\frac{e^2 aR}{(R^2 - a^2)^2}$$

Now, if $R \gg a$,

$$F \approx -\frac{e^2 a}{R^3} \propto \frac{1}{R^3}$$

If $R = a + \delta$, then

$$F \approx -\frac{e^2 a(a + \delta)}{(2a\delta)^2} \approx -\frac{e^2}{4\delta^2} \propto \frac{1}{\delta^2} = \frac{1}{(R-a)^2}$$

We may look at this case in a different way, for $R = a + \delta$, the sphere looks like an infinite conducting plane. So the force should not differ much from the force between the charge e and the image charge $-e$ (see Problem 3.9).

$$F = -\frac{e^2}{(2\delta)^2} = -\frac{e^2}{4\delta^2} \propto \frac{1}{\delta^2}$$

3.7 Conducting Cylinder and Line Charge (Stony Brook, Michigan State)

a) The image line charge together with the wire should provide a constant potential on the surface of the cylinder. The potential due to the image

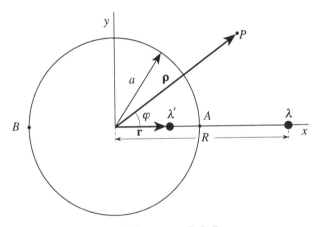

Figure   S.3.7

line charge at a distance $\boldsymbol{\rho}$ (point P in Figure S.3.7) is

$$\phi' = -2\lambda' \ln |\boldsymbol{\rho} - \mathbf{r}|$$

where λ' is the linear charge density and $\mathbf{r}$ is the distance from the axis of the cylinder to the line charge (see Figure S.3.7). The potential due to the charged wire, similarly, is

$$\phi = -2\lambda \ln |\boldsymbol{\rho} - R\hat{\mathbf{x}}|$$

From the condition that

$$\phi(\rho) = -2\lambda \ln |\boldsymbol{\rho} - R\hat{\mathbf{x}}| - 2\lambda' \ln |\boldsymbol{\rho} - \mathbf{r}| \qquad \text{(S.3.7.1)}$$

goes to 0 as $\rho \to \infty$, we have

$$\lambda' = -\lambda \qquad \text{(S.3.7.2)}$$

Again taking for simplicity two opposite points A and B on the surface of the cylinder, we find

$$-2\lambda \ln |a - R| - 2\lambda' \ln |a - r| = \phi_0 \qquad \text{(S.3.7.3)}$$

$$-2\lambda \ln |a + R| - 2\lambda' \ln |a + r| = \phi_0 \qquad \text{(S.3.7.4)}$$

Subtracting (S.3.7.4) from (S.3.7.3) and using (S.3.7.2), we obtain

$$\ln \left| \frac{a - R}{a + R} \right| = \ln \left| \frac{a - r}{a + r} \right| \qquad \text{(S.3.7.5)}$$

which yields the position of the image charge λ':

$$r = \frac{a^2}{R}$$

The potential on the surface

$$\phi = -2\lambda \ln |R - a| + 2\lambda \ln |a - a^2/R| = -2\lambda \ln \frac{R}{a}$$

We can check (see (b)) that this is the potential for any two points on the surface of the cylinder.

b) From (S.3.7.1) and (S.3.7.2), we have

$$\phi = -2\lambda \ln |\boldsymbol{\rho} - R\hat{\mathbf{x}}| + 2\lambda \ln |\boldsymbol{\rho} - r\hat{\mathbf{x}}|$$

In cylindrical coordinates, $\boldsymbol{\rho} = \rho \cos \varphi \hat{\mathbf{x}} + \rho \sin \varphi \hat{\mathbf{y}}$. So

$$\phi = 2\lambda \ln \left[\frac{\sqrt{(\rho \cos \varphi - r)^2 + \rho^2 \sin^2 \varphi}}{\sqrt{(\rho \cos \varphi - R)^2 + \rho^2 \sin^2 \varphi}} \right]$$

$$= \lambda \ln \left[\frac{\rho^2 - 2\rho(a^2/R) \cos \varphi + (a^4/R^2)}{\rho^2 - 2\rho R \cos \varphi + R^2} \right]$$

For $\rho = a$, $\phi = -2\lambda \ln(R/a)$, as in (a).

3.8   Spherical Void in Dielectric (Princeton)

We expect the dipole to induce some charge in the dielectric which would create a constant electric field inside the void, proportional to the dipole moment **p**. Therefore the field inside is due to the dipole field plus (presumably) a constant field. The field outside is the "screened" dipole field, which goes to zero at infinity. We look for a solution in the form

$$\mathbf{E}^{\text{in}} = \frac{3\mathbf{n}(\mathbf{n} \cdot \mathbf{p}) - \mathbf{p}}{r^3} + \alpha \mathbf{p} \qquad (S.3.8.1)$$

$$\mathbf{E}^{\text{out}} = \beta \frac{3\mathbf{n}(\mathbf{n} \cdot \mathbf{p}) - \mathbf{p}}{r^3} \qquad (S.3.8.2)$$

where **n** is normal to the surface of the void (see Figure S.3.8). Use the

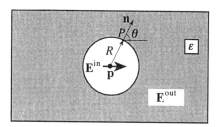

Figure   **S.3.8**

boundary conditions to find the coefficients α and β

$$D_{\text{n}}^{\text{in}} = D_{\text{n}}^{\text{out}}$$

$$E_{\text{t}}^{\text{in}} = E_{\text{t}}^{\text{out}}$$

and $\mathbf{D} = \varepsilon \mathbf{E}$, so

$$E_{\text{n}}^{\text{in}} = \varepsilon E_{\text{n}}^{\text{out}}$$

Write (S.3.8.1) and (S.3.8.2) at some point P on the surface of the void

$$\frac{2p \cos \theta}{R^3} + \alpha p \cos \theta = \varepsilon \beta \frac{2p \cos \theta}{R^3} \qquad (S.3.8.3)$$

$$-\frac{p \sin \theta}{R^3} + \alpha p \sin \theta = -\beta \frac{p \sin \theta}{R^3} \qquad (S.3.8.4)$$

where θ is the angle between **p** and the normal to the surface of the void,

$$\frac{2}{R^3} + \alpha = \frac{2\varepsilon \beta}{R^3} \qquad (S.3.8.5)$$

$$-\frac{1}{R^3} + \alpha = -\frac{\beta}{R^3} \qquad (S.3.8.6)$$

We solve for α and β in (S.3.8.5) and (S.3.8.6):

$$\beta = \frac{3}{2\varepsilon + 1} \tag{S.3.8.7}$$

$$\alpha = \frac{1 - \beta}{R^3} = \frac{2(\varepsilon - 1)}{(2\varepsilon + 1)R^3} \tag{S.3.8.8}$$

and find

$$\mathbf{E}^{in} = \frac{3\mathbf{n}(\mathbf{n} \cdot \mathbf{p}) - \mathbf{p}}{r^3} + \frac{2(\varepsilon - 1)}{2\varepsilon + 1} \frac{\mathbf{p}}{R^3} \tag{S.3.8.9}$$

$$\mathbf{E}^{out} = \frac{3}{2\varepsilon + 1} \frac{3\mathbf{n}(\mathbf{n} \cdot \mathbf{p}) - \mathbf{p}}{r^3} \tag{S.3.8.10}$$

3.9 Charge and Dielectric (Boston)

a) The electric fields just above and below the dielectric due to the charge e and the surface charge $\sigma_b(y, z)$ are, respectively,

$$\mathbf{E}(0^+, y, z) = \frac{e(-h\hat{\mathbf{x}} + y\hat{\mathbf{y}} + z\hat{\mathbf{z}})}{(h^2 + y^2 + z^2)^{3/2}} + 2\pi\sigma_b\hat{\mathbf{x}} \tag{S.3.9.1}$$

$$\mathbf{E}(0^-, y, z) = \frac{e(-h\hat{\mathbf{x}} + y\hat{\mathbf{y}} + z\hat{\mathbf{z}})}{(h^2 + y^2 + z^2)^{3/2}} - 2\pi\sigma_b\hat{\mathbf{x}} \tag{S.3.9.2}$$

b) From $\mathbf{D} = \mathbf{E} + 4\pi\mathbf{P} = \varepsilon\mathbf{E}$, we have for the polarization $\mathbf{P}$ and its normal component to the boundary P_n

$$\mathbf{P} = \frac{\varepsilon - 1}{4\pi}\mathbf{E}$$

$$\sigma_b = P_n = \frac{\varepsilon - 1}{4\pi}E_x(0^-, y, z) \tag{S.3.9.3}$$

c) Using (S.3.9.2) in (S.3.9.3), we have

$$\sigma_b = \frac{\varepsilon - 1}{4\pi}\left[-\frac{eh}{(h^2 + y^2 + z^2)^{3/2}} - 2\pi\sigma_b\right] \tag{S.3.9.4}$$

So

$$\sigma_b = -\frac{1}{2\pi}\frac{\varepsilon - 1}{\varepsilon + 1}\frac{eh}{(h^2 + y^2 + z^2)^{3/2}} \tag{S.3.9.5}$$

d) The field at the position of the charge e due to the surface charge σ_b is

$$\mathbf{E}'(h, 0, 0) = \int \frac{h\sigma_b \, dA}{(h^2 + y^2 + z^2)^{3/2}}\hat{\mathbf{x}} \tag{S.3.9.6}$$

where dA is the area. We can rewrite this integral in cylindrical coordinates:

$$\mathbf{E}'(h,0,0) = -\frac{e}{2\pi}\frac{\varepsilon-1}{\varepsilon+1}\int_0^{2\pi} d\varphi \int_0^\infty \frac{h^2\rho\,d\rho}{(h^2+\rho^2)^3}\,\hat{\mathbf{x}}$$

$$= -\frac{e}{2}\frac{\varepsilon-1}{\varepsilon+1}\int_0^\infty \frac{h^2 d\xi\,\hat{\mathbf{x}}}{(h^2+\xi)^3} = +\frac{e}{4}\frac{\varepsilon-1}{\varepsilon+1}\frac{h^2}{(h^2+\xi)^2}\bigg|_0^\infty \hat{\mathbf{x}} = -\frac{\varepsilon-1}{\varepsilon+1}\frac{e}{(2h)^2}\,\hat{\mathbf{x}}$$

$$(S.3.9.7)$$

This can be interpreted as an image charge

$$e' = -\frac{\varepsilon-1}{\varepsilon+1}e \qquad\qquad (S.3.9.8)$$

at a distance $2h$ from the charge e (see Figure S.3.9).

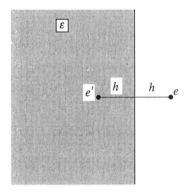

Figure   S.3.9

e) The force on the charge e is

$$\mathbf{F} = q\mathbf{E}'(h,0,0) = -\frac{\varepsilon-1}{\varepsilon+1}\frac{e^2}{4h^2}\,\hat{\mathbf{x}} \qquad\qquad (S.3.9.9)$$

3.10   Dielectric Cylinder in Uniform Electric Field (Princeton)

First solution: Introduce polar coordinates in the plane perpendicular to the axis of the cylinder (see Figure S.3.10). In the same manner as in

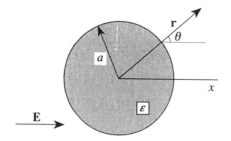

Figure   S.3.10

Problem 3.4, we will look for a potential outside the cylinder of the form

$$\phi^{\text{out}} = \phi_0 + \phi_1 \tag{S.3.10.1}$$

where $\phi_0 = -\mathbf{E} \cdot \mathbf{r}$ and ϕ_1 is a solution of the two-dimensional Laplace equation, which may depend on one constant vector $\mathbf{E}$

$$\phi_1 = A\mathbf{E} \cdot \nabla \ln r = A\frac{\mathbf{E} \cdot \mathbf{r}}{r^2} \tag{S.3.10.2}$$

where A is some constant (see, for instance, Landau and Lifshitz, *Electrodynamics of Continuous Media*, §8). Inside the cylinder, the only solution of Laplace's equation that is bounded in the center of the cylinder and depends on $\mathbf{E}$ is

$$\phi^{\text{in}} = B\mathbf{E} \cdot \mathbf{r}$$

Using the condition on the potential ϕ at $r = a$, $\phi^{\text{in}} = \phi^{\text{out}}$, we find

$$BE_0 a \cos \theta = -E_0 a \cos \theta + A\frac{E_0 a \cos \theta}{a^2} \tag{S.3.10.3}$$

from which we find

$$B = \frac{A}{a^2} - 1$$

We now have

$$\mathbf{E}^{\text{in}} = -\nabla \phi^{\text{in}} = \left(1 - \frac{A}{a^2}\right) E_0 \, \hat{\mathbf{x}} \tag{S.3.10.4}$$

$$\mathbf{E}^{\text{out}} = -\nabla \phi^{\text{out}} = \mathbf{E} - A\left(\frac{\mathbf{E}}{r^2} - 2\frac{(\mathbf{E} \cdot \mathbf{r}) \, \mathbf{r}}{r^4}\right) \tag{S.3.10.5}$$

Using the boundary condition $\mathbf{D}_{\text{n}}^{\text{in}} = \mathbf{D}_{\text{n}}^{\text{out}}$ or $\varepsilon \mathbf{E}_{\text{n}}^{\text{in}} = \mathbf{E}_{\text{n}}^{\text{out}}$, we find

$$\varepsilon \left(1 - \frac{A}{a^2}\right) E_0 \cos \theta = \left(1 + \frac{A}{a^2}\right) E_0 \cos \theta$$

$$A = \frac{\varepsilon - 1}{\varepsilon + 1} \, a^2$$

So we obtain

$$\mathbf{E}^{\text{in}} = \frac{2}{1 + \varepsilon} E_0 \hat{\mathbf{x}} \qquad (S.3.10.6)$$

$$\mathbf{E}^{\text{out}} = \mathbf{E} - A \left(\frac{\mathbf{E}}{r^2} - \frac{2(\mathbf{E} \cdot \mathbf{r})\mathbf{r}}{r^4} \right)$$

$$= E_0 \hat{\mathbf{x}} - \frac{\varepsilon - 1}{\varepsilon + 1} \, a^2 \left(\frac{E_0 \hat{\mathbf{x}}}{r^2} - \frac{2E_0 \cos \theta \, \mathbf{r}}{r^3} \right)$$

$$= \left(1 - \frac{\varepsilon - 1}{\varepsilon + 1} \frac{a^2}{r^2} \right) E_0 \hat{\mathbf{x}} + \frac{2a^2}{r^3} \frac{\varepsilon - 1}{\varepsilon + 1} E_0 \cos \theta \, \mathbf{r} \qquad (S.3.10.7)$$

The polarization is

$$\mathbf{P} = \frac{\varepsilon - 1}{4\pi} \mathbf{E}^{\text{in}} = \frac{1}{2\pi} \frac{\varepsilon - 1}{\varepsilon + 1} E_0 \hat{\mathbf{x}}$$

So the dipole moment per unit length of the cylinder is

$$\mathbf{p} = \mathbf{P} \cdot \pi a^2 = \frac{1}{2} \frac{\varepsilon - 1}{\varepsilon + 1} \, a^2 E_0 \, \hat{\mathbf{x}}$$

which corresponds to the potential

$$\phi_1 = \frac{2\mathbf{p} \cdot \mathbf{r}}{r^2}$$

The surface charge density σ is

$$\sigma = \frac{\mathbf{P} \cdot \mathbf{r}}{r} = \frac{1}{2\pi} \frac{\varepsilon - 1}{\varepsilon + 1} E_0 \cos \theta$$

Second solution: Use the fact that for any dielectric ellipsoid with a dielectric constant ε immersed in a uniform electric field in vacuum, a uniform electric field inside is created (see, for instance, Landau and Lifshitz, *Electrodynamics of Continuous Media*, §8). Therefore there must be a linear dependence between E_{0x}, E_x^{in}, and D_x^{in}, where the applied field E_0 is along the x-axis:

$$E_0 = E_{0x} = \alpha E_x^{\text{in}} + \beta D_x^{\text{in}} \qquad (S.3.10.8)$$

where α and β are coefficients independent of the dielectric constant of the ellipsoid and only depend on its shape. For the trivial case in which $\varepsilon = 1$

$$E_0 = E_x^{\text{in}} = D_x^{\text{in}}$$

Therefore

$$\alpha + \beta = 1 \qquad (S.3.10.9)$$

For a conducting ellipsoid (which can be described by a dielectric constant $\varepsilon = \infty$)

$$E_x^{\text{in}} = 0$$

$$\beta = \frac{E_0}{D_x^{\text{in}}} = \frac{E_0}{4\pi P_x} = n_x$$

where n_x is the depolarization factor. From (S.3.10.9), we have

$$\alpha = 1 - n_x$$

Finally (S.3.10.8) takes the form

$$(1 - n_x)\, E_x^{\text{in}} + n_x D_x^{\text{in}} = E_0 \qquad (S.3.10.10)$$

For a cylinder parallel to the applied field along the z-axis, $n_z = 0$, but $n_x + n_y + n_z = 1$, so $n_x = n_y = 1/2$. Equation (S.3.10.10) becomes

$$\frac{1}{2} E_x^{\text{in}} + \frac{1}{2}\varepsilon E_x^{\text{in}} = E_0 \qquad (S.3.10.11)$$

and

$$E_x^{\text{in}} = \frac{2}{1 + \varepsilon} E_0 \qquad (S.3.10.12)$$

as in (S.3.10.6) above.

3.11 Powder of Dielectric Spheres (Stony Brook)

To find the effective dielectric constant, we must first find the polarization of the spherical particles. Consider a dielectric sphere placed in a uniform electric field $\mathbf{E} = E_0\hat{\mathbf{z}}$. Since the field at infinity is $\mathbf{E} = E_0\hat{\mathbf{z}}$ and the field produced by the sphere must be a dipole field (see Problem 3.10), try a solution outside the sphere:

$$\mathbf{E}_{\text{out}} = E_0\hat{\mathbf{z}} + \frac{3(\mathbf{p} \cdot \mathbf{r})\mathbf{r} - \mathbf{p}r^2}{r^5} \qquad (S.3.11.1)$$

Inside the sphere, try a field oriented in the direction of the original field E_0:

$$\mathbf{E}_{\text{in}} = E_1\hat{\mathbf{z}} \qquad (S.3.11.2)$$

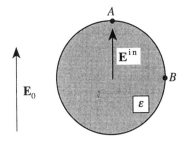

Figure   S.3.11

The usual boundary conditions apply:

$$D_n^{in} = D_n^{out} \qquad (S.3.11.3)$$

$$E_t^{in} = E_t^{out} \qquad (S.3.11.4)$$

Taking (S.3.11.3) and (S.3.11.4) at points A and B, respectively (see Figure S.3.11),

$$\varepsilon E_1 = E_0 + \frac{2p}{R^3} \qquad (S.3.11.5)$$

$$E_1 = E_0 - \frac{p}{R^3} \qquad (S.3.11.6)$$

Solving (S.3.11.5) and (S.3.11.6) for E_1 yields

$$E_1 = \frac{3E_0}{\varepsilon + 2} \qquad (S.3.11.7)$$

so

$$\mathbf{E}_{in} = \frac{3E_0}{\varepsilon + 2}\hat{\mathbf{z}}$$

The dipole moment p is found by substituting (S.3.11.7) back into (S.3.11.6):

$$\frac{3E_0}{\varepsilon + 2} = E_0 - \frac{p}{R^3}$$

$$p = \left(\frac{\varepsilon - 1}{\varepsilon + 2}\right) E_0 R^3 = \frac{3}{4\pi}\left(\frac{\varepsilon - 1}{\varepsilon + 2}\right) E_0 V$$

Using the condition $nR^3 \ll 1$ (low concentration of particles), we can disregard the interaction between them. The polarization of the medium then is given by the dipole moment per unit volume. Here, we have n dipoles per unit volume, so

$$P = \left(\frac{\varepsilon - 1}{\varepsilon + 2}\right) E_0 n R^3 \qquad (S.3.11.8)$$

Now, $D = \varepsilon' E_0 = E_0 + 4\pi P$, so

$$\varepsilon' = 1 + \frac{4\pi P}{E_0} = 1 + 4\pi \left(\frac{\varepsilon - 1}{\varepsilon + 2}\right) nR^3$$

$$= 1 + 4\pi \left(\frac{4 - 1}{4 + 2}\right) \left(\frac{10^{12}}{\text{cm}^3}\right) \left(10^{-5} \text{ cm}\right)^3 = 1.006 \qquad \text{(S.3.11.9)}$$

The apparent but wrong answer $\varepsilon' = 1 + (\varepsilon - 1)nV$ comes about by neglecting the shape of the particles and considering ε' as the average of the dielectric constant of free space, 1, and the dielectric constant of the spheres, ε.

3.12   Concentric Spherical Capacitor (Stony Brook)

a) By using Gauss's theorem and the fact that charges rearrange themselves so as to yield a zero electric field inside the conductors, we infer that all of Q_1 will reside on surface 2 of the inner sphere, $-Q_1$ on surface 3 of the outer sphere (no field in the interior of the outer sphere), and $Q_1 + Q_2$ on surface 4 of the outer sphere (see Figure S.3.12). The surface charge

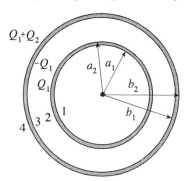

Figure   S.3.12

densities are straightforward to calculate as the charge divided by surface area:

$$\sigma_1 = 0 \qquad \sigma_2 = \frac{Q_1}{4\pi a_2^2} \qquad \sigma_3 = \frac{-Q_1}{4\pi b_1^2} \qquad \sigma_4 = \frac{Q_1 + Q_2}{4\pi b_2^2}$$

If $Q_2 = -Q_1$, there is no charge on the external shell, simply Q_1 on surface 2 and $-Q_1$ on surface 3. The mutual capacitance may be calculated from $\Delta V C = Q_1$, where ΔV is the difference in electric potential between the spherical shells. Again using Gauss's theorem to calculate the magnitude

E of the electric field between the shells, we have

$$E(r) \cdot 4\pi r^2 = 4\pi Q_1 \qquad\qquad E(r) = \frac{Q_1}{r^2}$$

$$-\Delta V = \int_{a_2}^{b_1} E(r) \cdot dr = -Q_1 \left(\frac{1}{b_1} - \frac{1}{a_2} \right)$$

$$\frac{1}{C} = \frac{1}{a_2} - \frac{1}{b_1}$$

b) The $\mathbf{D}$ field behaves like the $\mathbf{E}$ field before:

$$\nabla \cdot \mathbf{D} = 4\pi\rho$$

so, between the spheres

$$\mathbf{D} = \varepsilon \mathbf{E} = \frac{Q_1}{r^2}$$

$$\mathbf{E} = \frac{Q_1}{\varepsilon r^2}$$

The real and polarization surface charge densities on surface 1 are still zero, and on surface 2, we find

$$\sigma^{(2)} = \frac{Q_1}{4\pi a_2^2}$$

$$\sigma^{(2)}_{\text{pol}} = -\frac{Q_1}{4\pi a_2^2} \frac{\varepsilon - 1}{\varepsilon}$$

Likewise, on the third and fourth surfaces,

$$\sigma^{(3)}_{\text{pol}} = \frac{+Q_1}{4\pi b_1^2} \frac{\varepsilon - 1}{\varepsilon}$$

$$\sigma^{(4)}_{\text{pol}} = 0$$

Finally, the capacitance may be found:

$$-\Delta V = \int_{a_2}^{b_1} E \cdot dr = -\frac{Q_1}{\varepsilon} \left(\frac{1}{b_1} - \frac{1}{a_2} \right)$$

$$C = \varepsilon \left(\frac{1}{a_2} - \frac{1}{b_1} \right)^{-1}$$

3.13   Not-so-concentric Spherical Capacitor (Michigan Tech)

a) For $\delta = 0$, we have the boundary conditions

$$\phi|_{r=b} = 0 \qquad \phi|_{r=a} = \text{const} \qquad \text{(S.3.13.1)}$$

which with Gauss's law yield for the potential between a and b

$$\phi = q\left(\frac{1}{r} - \frac{1}{b}\right) \qquad \text{(S.3.13.2)}$$

b) Introduce spherical coordinates with the polar axis along the line $\overline{AB}$ (see Figure S.3.13). Find the equation of the sphere S_B in these coordinates.

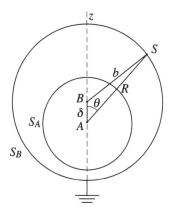

Figure S.3.13

From the triangle $\triangle ABS$ we have

$$b^2 = R^2 + \delta^2 - 2R\delta\cos\theta$$

We can expand $1/b$ as a sum of spherical harmonics using a general formula:

$$\frac{1}{b} = \frac{1}{\sqrt{R^2 + \delta^2 - 2\delta R\cos\theta}} = \sum_{l=0}^{\infty}\frac{\delta^l}{R^{l+1}}P_l(\cos\theta)$$

$$\approx \frac{1}{R} + \frac{\delta}{R^2}P_1(\cos\theta) = \frac{1}{R} + \frac{\delta}{R^2}\cos\theta$$

or simply by expanding the square root to first order in δ ($\delta \ll R$)

$$\frac{1}{b} \approx \frac{1}{R}\left(1 + \frac{1}{2}\,2\frac{\delta}{R}\cos\theta\right)$$

So

$$b \approx R \left(1 - \frac{\delta}{R} \cos\theta \right) = R - \delta\cos\theta$$

and we have

$$R \approx b + \delta\cos\theta \qquad (S.3.13.3)$$

The term $\delta P_1(\cos\theta)$ represents the deviation from concentricity and should be zero at $\delta = 0$. We look for a potential as an expansion of spherical harmonics to first order in δ

$$\phi(r,\theta) = \left(A_0 + \frac{B_0}{r} \right) + \delta \left(A_1 r + \frac{B_1}{r^2} \right) \cos\theta \qquad (S.3.13.4)$$

With the boundary conditions in (S.3.13.1)

$$\phi|_{r=R} = 0 \qquad\qquad \phi|_{r=a} = \text{const}$$

The first term in (S.3.13.4) should be the same as in (a)

$$\phi = q \left(\frac{1}{r} - \frac{1}{b} \right)$$

We may find A_1 and B_1 by checking the potential on the inner and outer spheres. On the inner sphere $r = a$ and the potential is a constant, and so must be independent of $\cos\theta$. This yields

$$A_1 a + \frac{B_1}{a^2} = 0$$
$$\qquad\qquad\qquad\qquad\qquad (S.3.13.5)$$
$$B_1 = -a^3 A_1$$

Substituting (S.3.13.5) back into (S.3.13.4), we now check the potential on the outer sphere, where $r = b + \delta\cos\theta$

$$q \left(\frac{1}{\delta\cos\theta + b} - \frac{1}{b} \right) + \delta A_1 \left(b + \delta\cos\theta - \frac{a^3}{(b + \delta\cos\theta)^2} \right) \cos\theta = 0$$

$$\approx q \left[\frac{1}{b} \left(1 - \frac{\delta}{b}\cos\theta \right) - \frac{1}{b} \right] + \delta A_1 \left[b + \delta\cos\theta - \frac{a^3}{b^2} \left(1 - \frac{2\delta\cos\theta}{b} \right) \right] \cos\theta$$

$$= q \left(-\frac{\delta\cos\theta}{b^2} \right) + \delta A_1 \left[b + \delta\cos\theta - \frac{a^3}{b^2} \left(1 - \frac{2\delta\cos\theta}{b} \right) \right] \cos\theta$$

Neglecting terms of order δ^2, we find

$$\delta \cos\theta \left(-\frac{q}{b^2} + A_1 b - \frac{A_1 a^3}{b^2} \right) = 0$$

$$A_1 = \frac{q}{b^3 - a^3}$$

Finally,

$$\phi(r,\theta) = q \left(\frac{1}{r} - \frac{1}{b} \right) + \frac{\delta q}{b^3 - a^3} \left(r - \frac{a^3}{r^2} \right) \cos\theta$$

The charge density on the inner sphere is

$$\sigma = -\frac{1}{4\pi} \left. \frac{\partial\phi}{\partial n} \right|_{r=a} = \frac{q}{4\pi a^2} - \frac{3q\delta}{4\pi (b^3 - a^3)} \cos\theta$$

The force on the sphere may now be calculated by integrating the z component of the force on the differential areas of the surface $d\mathbf{F} = 2\pi\sigma^2\mathbf{n}\,dA$:

$$F = F_z = \int_0^\pi 2\pi a^2 \cdot 2\pi\sigma^2 \sin\theta\,d\theta\ \cos\theta$$

$$= \int_0^\pi 4\pi^2 a^2 \left(\frac{q}{4\pi a^2} - \frac{3q\delta}{4\pi (b^3 - a^3)} \cos\theta \right)^2 \sin\theta \cos\theta\,d\theta$$

The only term which survives is the cross term

$$F = \int_0^\pi 4\pi^2 a^2 \left(-2\frac{3q^2\delta \cos\theta}{16\pi^2 a^2 (b^3 - a^3)} \right) \sin\theta\ \cos\theta\,d\theta$$

$$= -\frac{q^2\delta}{(b^3 - a^3)} \tag{S.3.13.6}$$

We can check this result in the limit of $a = 0$ against the force between a charge inside a neutral sphere and the sphere (see Problems 3.5, 3.6).

3.14 Parallel Plate Capacitor with Solid Dielectric (Stony Brook, Michigan Tech, Michigan)

To solve this problem, the capacitance of a parallel plate capacitor must be calculated with and without a dielectric inserted. We then recognize that the two capacitors are in parallel in order to infer the total capacitance, and we then use the expression for the energy contained in the capacitor to determine the force on the dielectric. The potential difference between the plates is given by

$$V = - \int_0^d \mathbf{E} \cdot d\mathbf{l} = 4\pi\sigma d \qquad (\text{S.3.14.1})$$

where σ is the surface charge density $\sigma = Q/L^2$. Since $Q = CV$, we find

$$C = \frac{\sigma L^2}{4\pi\sigma d} = \frac{L^2}{4\pi d} \qquad (\text{S.3.14.2})$$

a) With a dielectric inserted, the capacitance is modified. $\mathbf{E}$ is replaced by $\mathbf{D}$ in (S.3.14.1), where $\mathbf{D} = \varepsilon\mathbf{E}$. The potential V is still the integral of $-\mathbf{E} \cdot d\mathbf{l}$, so we find that the capacitance is multiplied by ε. In this problem, the dielectric is inserted only a distance x between the plates (see Figure

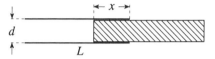

Figure S.3.14

S.3.14). As the total capacitance of two capacitors in parallel is simply the sum of the individual capacitances, we find that the new capacitance is

$$C' = \frac{L}{4\pi d}(\varepsilon x + L - x) \qquad (\text{S.3.14.3})$$

The energy stored in a capacitor is given by

$$U = \frac{1}{2}\frac{Q^2}{C} = \frac{1}{2}CV^2 \qquad (\text{S.3.14.4})$$

In (a), the battery has been disconnected from the capacitor. As the dielectric moves either into or out of the capacitor, the potential V will change while the charge Q is constant. So we use the first part of (S.3.14.4). The force on the dielectric is found from

$$F_x = -\frac{\partial U}{\partial x} \qquad (\text{S.3.14.5})$$

From (S.3.14.3) and (S.3.14.4) we have

$$U = \frac{1}{2}\frac{4\pi d Q^2}{L}\left(\frac{1}{L + x(\varepsilon - 1)}\right)$$

so

$$F_x = \frac{2\pi d Q^2}{L} \frac{(\varepsilon - 1)}{[L + x(\varepsilon - 1)]^2} \qquad \text{(S.3.14.6)}$$

Expressing F_x in terms of a potential difference V

$$Q = C'V$$

we obtain

$$F_x = \frac{LV^2}{8\pi d}(\varepsilon - 1)$$

Since $\varepsilon > 1$, the dielectric slab will be drawn further between the plates (x will increase).

b) In this case, V is constant, not Q, and we must now take into account the electric potential energy of the battery. For some small change of the system,

$$\Delta U = \Delta \left(\frac{1}{2} C V^2 \right) - V \Delta Q$$

Since $Q = CV$, and the potential is held fixed, $\Delta Q = V \Delta C$, so

$$\Delta U = -\frac{1}{2} V^2 \Delta C$$

Note the minus sign. If the electric potential energy of the battery had been ignored, the expression would be incorrect (see also Problem 3.17). We now wish to find

$$F_x = -\frac{\partial U}{\partial x} = +\frac{V^2}{2} \frac{\partial C'}{\partial x} = \frac{LV^2}{8\pi d}(\varepsilon - 1) \qquad \text{(S.3.14.7)}$$

as obtained in (a). Note that the force goes to zero when $\varepsilon = 1$, as expected.

3.15   Parallel Plate Capacitor in Dielectric Bath (MIT)

a) As in Problem 3.14, the capacitance of the parallel plate and dielectric system is simply the sum of two capacitors in parallel:

$$C = \frac{L(L/2)}{4\pi d} + \frac{\varepsilon L(L/2)}{4\pi d} = \frac{L^2}{8\pi d}(1 + \varepsilon) \equiv C_0 \frac{1 + \varepsilon}{2}$$

b) The charge is constant as the plates are lowered into the dielectric bath, but the potential between the plates is not. After charging, $Q = C_0 V$. Lowering the plates into the fluid to a height $L/2$ changes C from C_0 to

$C_0(1 + \varepsilon)/2$. The new potential V' may be found from

$$C_0 V = C_0 \frac{1 + \varepsilon}{2} V'$$

So

$$V' = \frac{2}{1 + \varepsilon} V$$

Since $|V| = |\mathbf{E}|\, d$

$$E = \frac{2}{1 + \varepsilon} \frac{V}{d}$$

c) From Maxwell's equation $\nabla \cdot \mathbf{D} = 4\pi\rho$, we find that $D = 4\pi\sigma$, but $D = \varepsilon E$, so

$$E = \frac{4\pi\sigma}{\varepsilon}$$

From (b), we obtain

$$\sigma = \frac{2\varepsilon}{4\pi(1 + \varepsilon)} \frac{V}{d}$$

where the fluid is between the plates. The surface charge density where there is no fluid is

$$\sigma = \frac{2}{4\pi(1 + \varepsilon)} \frac{V}{d}$$

d) To determine the height difference between the liquid between the plates and in the external reservoir, we consider the sum of the electrical and gravitational potential energies of the capacitor and fluid. Let the height difference be given by h and a small change in the height be produced by ζ (see Figure S.3.15) . The gravitational potential energy is given by the integral

$$\int_0^{h+\zeta} \rho g L d\, x\, dx = \rho g \frac{(h+\zeta)^2 L d}{2}$$

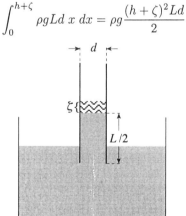

Figure S.3.15

so that the total potential energy

$$U = \frac{1}{2}\frac{Q^2}{C} + \frac{\rho g(h+\zeta)^2 Ld}{2} \qquad (S.3.15.1)$$

Again we have written the electrical potential energy in terms of the charge Q, since the potential changes as the fluid rises or falls. Writing out the capacitance, we get

$$U = \frac{1}{2}\frac{8\pi Q^2 d}{L^2(\varepsilon+1) + 2L(\varepsilon-1)\zeta} + \frac{\rho g(h+\zeta)^2 Ld}{2} \qquad (S.3.15.2)$$

At equilibrium, the force on the liquid is zero, or the derivative of the potential energy is zero

$$-\frac{\partial U}{\partial \zeta} = \frac{1}{2}\frac{16\pi Q^2 Ld(\varepsilon-1)}{[L^2(\varepsilon+1) + 2L(\varepsilon-1)\zeta]^2} - \rho g(h+\zeta)Ld = 0$$

At equilibrium, i.e., $\zeta = 0$,

$$\frac{8\pi Q^2 Ld(\varepsilon-1)}{L^4(\varepsilon+1)^2} = \rho g Ldh \qquad (S.3.15.3)$$

$$h = \frac{8\pi Q^2(\varepsilon-1)}{L^4 \rho g(\varepsilon+1)^2} \qquad (S.3.15.4)$$

Rewriting (S.3.15.4) in terms of V, we obtain

$$h = \frac{V^2}{2\pi \rho g d^2}\frac{\varepsilon-1}{(\varepsilon+1)^2}$$

Alternatively, we can use the result of Problem 3.14 (S.3.14.6) and equate the force

$$F_h = \left(-\frac{\partial U}{\partial x}\right)_{x=L/2} = \frac{2\pi dQ^2}{L}\frac{(\varepsilon-1)}{[L + L(\varepsilon-1)/2]^2} \qquad (S.3.15.5)$$

to the weight of the dielectric

$$W = \rho g h Ld \qquad (S.3.15.6)$$

From (S.3.15.5) and (S.3.15.6) we find

$$h = \frac{8\pi Q^2}{L^4 \rho g}\frac{(\varepsilon-1)}{(\varepsilon+1)^2}$$

as in (S.3.15.4).

3.16   Not-so-parallel Plate Capacitor (Princeton (a), Rutgers (b))

a) Consider this problem in cylindrical coordinates, so that the plates are along the radii (see Figure S.3.16). The Laplace equation, $\nabla^2\phi = 0$, then

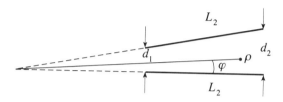

Figure S.3.16

becomes

$$\frac{1}{\rho}\frac{\partial}{\partial\rho}\left(\rho\frac{\partial\phi}{\partial\rho}\right) + \frac{1}{\rho^2}\frac{\partial^2\phi}{\partial\varphi^2} = 0 \qquad (S.3.16.1)$$

Writing $\phi = R(\rho)\chi(\varphi)$, we may separate (S.3.16.1) into two differential equations

$$\frac{\rho}{R}\frac{d}{d\rho}\left(\rho\frac{dR}{d\rho}\right) = n^2 \qquad (S.3.16.2)$$

$$\frac{1}{\chi}\frac{d^2\chi}{d\varphi^2} = -n^2 \qquad (S.3.16.3)$$

where R is the radial part of the potential, χ is the azimuthal part, and n is some constant. In the small-angle approximation (which can be assumed since we are allowed to disregard the edge effects), we can say that R is independent of ρ, and then

$$\frac{d^2\chi}{d\varphi^2} = 0$$

for which we have the solution $\chi = A\varphi + B$. From the boundary condition $\chi(0) = 0$, we have $B = 0$, and using the other condition $\chi(\varphi_0) = V$, we find that

$$\phi = V\frac{\varphi}{\varphi_0}$$

where

$$\varphi_0 \approx \sin\varphi_0 = \frac{d_2 - d_1}{L_2} \qquad (S.3.16.4)$$

b) Using the result of (a) and the expressions for the of the field energy U contained in a capacitor, we may find the capacitance. We have

$$U = \frac{1}{2}CV^2 = \frac{1}{8\pi}\int E^2\, d^3x \qquad (S.3.16.5)$$

Here,

$$\mathbf{E} = -\nabla\phi = -\frac{V}{\rho\varphi_0}\mathbf{e}_\varphi$$

so (S.3.16.5) becomes

$$\frac{1}{2}CV^2 = \frac{1}{8\pi}\int_0^{L_1}\int_{\rho_1}^{\rho_2}\int_0^{\varphi_0}\frac{V^2}{\rho^2\varphi_0^2}\rho\, d\varphi\, d\rho\, dz$$

$$= \frac{1}{8\pi}\frac{L_1 V^2}{\varphi_0}\int_{\rho_1}^{\rho_2}\frac{d\rho}{\rho} = \frac{L_1 L_2 V^2}{8\pi\,(d_2 - d_1)}\ln\frac{d_2/\varphi_0}{d_1/\varphi_0} = \frac{L_1 L_2 V^2}{8\pi(d_2 - d_1)}\ln\frac{d_2}{d_1}$$

Hence, the capacitance

$$C = \frac{L_1 L_2}{4\pi(d_2 - d_1)}\ln\frac{d_2}{d_1} = \frac{L_1 L_2}{4\pi(d_2 - d_1)}\ln\left(1 + \frac{d_2 - d_1}{d_1}\right) \qquad (S.3.16.6)$$

In the limit of $d_2 \to d_1 = d$ (the case of a parallel plate capacitor), (S.3.16.6) reduces to

$$C \approx \frac{L_1 L_2}{4\pi\,(d_2 - d_1)}\cdot\frac{d_2 - d_1}{d_1} = \frac{L_1 L_2}{4\pi d}$$

which is equivalent to the result found in (S.3.14.2) of Problem 3.14.

3.17 Cylindrical Capacitor in Dielectric Bath (Boston, Maryland)

a) For the first case (fixed charge), the generalized force can be calculated as usual by considering the change in potential energy of the capacitor (field source) written in terms of the charge of the capacitor (which is a closed system). So, in this case,

$$F_{ex} = -\left(\frac{\partial U_e}{\partial x}\right)_Q \qquad (S.3.17.1)$$

If the capacitor is connected to the battery, it is no longer a closed system, and we have to consider the energy of the battery also. The battery must do some work to keep the potential of the plate constant. This work ΔW

is

$$\Delta W = -V_b \Delta Q = -V_b^2 \Delta C = -2\Delta U_e$$

The energy change of the total system

$$\Delta \tilde{U}_e = \Delta U_e + \Delta W = -\Delta U_e$$

Therefore, the force

$$F_{ex} = -\left(\frac{\partial \tilde{U}_e}{\partial x}\right)_V = +\frac{dU_e(x, V_b)}{dx} \qquad (S.3.17.2)$$

b) The energy of the capacitor

$$U_e = \frac{Q^2}{2C} = \frac{2\pi Q^2 x}{A} = \frac{AV_b^2}{8\pi x}$$

where A is the area of the plates. From (S.3.17.1)

$$F_{ex} = -\frac{dU_e(x, Q)}{dx} = -\frac{2\pi Q^2}{A}$$

In the case of constant voltage

$$F_{ex} = +\frac{\partial U_e(x, V_b)}{\partial x} = -\frac{AV_b^2}{8\pi x^2} = -\frac{2\pi Q^2}{A}$$

c) The capacitance of a cylindrical capacitor may be found by calculating the potential outside a uniformly charged cylinder. Gauss's theorem gives

$$\phi(r) = 2\lambda \ln(r/a) \qquad (S.3.17.3)$$

where λ is the linear charge density of the cylinder, and a is the radius of the cylinder. The potential of the outer cylinder of the capacitor in the problem is V. So

$$\phi(b) = V = 2\lambda \ln(b/a) \qquad (S.3.17.4)$$

For a cylinder of length H, the capacitance may be found from (S.3.17.4) by substituting $\lambda = Q/H$:

$$V = \frac{Q}{C} = \frac{2Q \ln(b/a)}{H} \qquad (S.3.17.5)$$

$$C = \frac{H}{2 \ln(b/a)} \qquad (S.3.17.6)$$

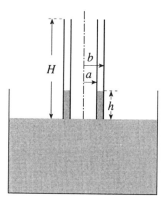

For the capacitor of length H filled with dielectric up to a height h (see Figure S.3.17)

$$U_e = \frac{1}{2} \frac{1}{2\ln(b/a)} \left[(H - h) + \varepsilon h\right] V^2$$

Here, we use (S.3.17.2) to obtain

$$F_e = \frac{\partial U_e}{\partial h} = \frac{\varepsilon - 1}{4\ln(b/a)} V^2 \qquad\qquad (S.3.17.7)$$

The liquid is drawn into the capacitor. The weight of the liquid in the capacitor

$$W = \pi \left(b^2 - a^2\right) h\rho g \qquad\qquad (S.3.17.8)$$

where $\pi(b^2 - a^2)h$ is the volume of the liquid drawn up between the cylinders and ρ is the mass density of the liquid. Equating (S.3.17.7) and (S.3.17.8), we get

$$h = \frac{(\varepsilon - 1)V^2}{4\pi\rho g \left(b^2 - a^2\right) \ln(b/a)}$$

3.18   Iterated Capacitance (Stony Brook)

a) We found in Problem 3.5 that a charge q a distance R from a conducting sphere of radius a produced an image charge q' a distance r from the center of the sphere, where

$$q' = -\frac{qa}{R} \qquad\qquad r = \frac{a^2}{R} \qquad\qquad (S.3.18.1)$$

Using this result, we may verify that charges q and q' a distance $R - r$

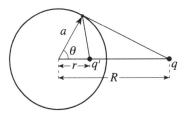

Figure   S.3.18a

apart do indeed give a spherical equipotential surface. The problem is
cylindrically symmetric, so we establish a circular equipotential by writing
down the potential due to the two charges (see Figure S.3.18a). Given that
$V(\infty) = 0$, we find the sum of the potentials

$$\frac{q}{|\mathbf{R} - \mathbf{a}|} + \frac{q'}{|\mathbf{r} - \mathbf{a}|} = \frac{q}{\sqrt{R^2 + a^2 - 2aR\cos\theta}} + \frac{q'}{\sqrt{a^2 + r^2 - 2ar\cos\theta}}$$

for a circle of radius a. If $q' = -qa/R$, we have

$$\frac{q}{\sqrt{R^2 + a^2 - 2aR\cos\theta}} - \frac{q}{(R/a)\sqrt{a^2 + r^2 - 2ar\cos\theta}} = 0$$

With $r = a^2/R$,

$$\frac{q}{\sqrt{R^2 + a^2 - 2aR\cos\theta}} - \frac{q}{\sqrt{R^2 + a^2 - 2aR\cos\theta}} = 0$$

as required. Given that the separation of the two charges is some distance
$x = R - r$, we may find the radius of the sphere a and the location of the
center with respect to one of the charges R using (S.3.18.1):

$$a = -\frac{q'}{q}\frac{x}{1 - (q'/q)^2} \qquad\qquad R = \frac{x}{1 - (q'/q)^2}$$

b),c) In general, the charge and potentials of a number of conductors are
related by the linear equations

$$q_i = \sum_{j=1}^{n} C_{ij}V_j \qquad\qquad (\text{S.3.18.2})$$

where C_{ii} and C_{ij} are called coefficients of capacity and induction, respectively. In the case of two conductors carrying equal but opposite charges,
the capacitance is defined by the ratio of the charge on one conductor to
the potential difference between them. For our two-sphere capacitor, we
have

$$q_1 = C_{11}V_1 + C_{12}V_2$$

$$q_2 = C_{21}V_1 + C_{22}V_2 \qquad\qquad (\text{S.3.18.3})$$

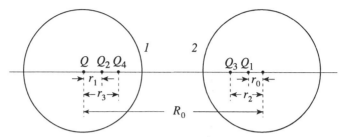

Figure   S.3.18b

The capacitance may be found by setting $q_2 = -q_1$ and calculating

$$C = \frac{q_1}{V_1 - V_2} \tag{S.3.18.4}$$

If we choose the zero of potential at infinity, it is clear that $V_2 = -V_1$, so we have for the capacitance

$$C = \frac{q_1}{V_1 - V_2} = \frac{q_1}{2V_1} = \frac{1}{2}(C_{11} - C_{12}) \tag{S.3.18.5}$$

This is a specific case of a more general result (see, for instance, Landau and Lifshiftz, *Electrodynamics of Continuous Media*, §2)

$$C = \frac{C_{11}C_{22} - C_{12}^2}{C_{11} + 2C_{12} + C_{22}} \tag{S.3.18.6}$$

Now we may calculate C_{11} and C_{12} by placing a charge Q on the first sphere, giving it a potential $V = Q/a$, then placing an image charge inside the other sphere to keep it at zero potential. In turn, we place an image charge inside the first sphere to restore the potential to Q/a, whereupon we. ... Operationally, we have from (S.3.18.1) (see Figure S.3.18b)

$$Q_0 = Q$$

$$R_0 = R \qquad Q_1 = -\frac{Q_0 a}{R_0} \qquad r_0 = \frac{a^2}{R_0}$$

$$R_{i-1} = R_0 - r_{i-2} \qquad Q_i = -\frac{Q_{i-1}a}{R_{i-1}} \qquad r_{i-1} = \frac{a^2}{R_{i-1}}$$

c) For our problem, $R = 4a$, and $Q_0 = Q$, so we find $Q_1 = -Q/4$, $Q_2 = Q/15$, $Q_3 = -Q/56$, and $Q_4 \approx Q/209$. Since we maintain $V_2 = 0$, we have from the first of (S.3.18.3)

$$q_1 = C_{11}V_1 \approx Q + Q_2 + Q_4 = C_{11}\frac{Q}{a} \tag{S.3.18.7}$$

So

$$C_{11} \approx \left(1 + \frac{1}{15} + \frac{1}{209}\right) a \approx 1.07a \qquad (S.3.18.8)$$

Since $C_{12} = C_{21}$, we find from the second of (S.3.18.3) by summing the first two image charges in sphere 2

$$q_2 = C_{21}V_1 = C_{12}\frac{Q}{a} \approx Q_1 + Q_3 \qquad (S.3.18.9)$$

So

$$C_{12} \approx \left(-\frac{1}{4} - \frac{1}{56}\right) a \approx -0.27a \qquad (S.3.18.10)$$

Finally, we have

$$C = \frac{1}{2}\left(C_{11} - C_{12}\right) \approx \frac{1}{2}(1.07 + 0.27)a = 0.67a \qquad (S.3.18.11)$$

See Smythe, *Static and Dynamic Electricity* §5.08 for further details.

3.19   Resistance vs. Capacitance (Boston, Rutgers (a))

a) Enclose one of the conductors in a surface and use Maxwell's equation (see Figures S.3.19a and S.3.19b)

$$\nabla \cdot \mathbf{D} = 4\pi\rho$$

So

$$\nabla \cdot \mathbf{E} = \frac{4\pi\rho}{\varepsilon} \qquad (S.3.19.1)$$

Take the volume integral of (S.3.19.1) and transform into a surface integral:

$$\int \nabla \cdot \mathbf{E}\, d^3x = \int \frac{4\pi\rho}{\varepsilon}\, d^3x = \int \mathbf{E}\cdot d\mathbf{A} \qquad (S.3.19.2)$$

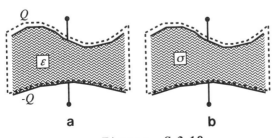

a   **b**

Figure   S.3.19

Using this result and the definition of C, we find

$$Q = \int \rho \, d^3 x = CV$$

which yields

$$\frac{1}{V} \int \mathbf{E} \cdot d\mathbf{A} = \frac{4\pi C}{\varepsilon} \qquad \text{(S.3.19.3)}$$

Now treat the resistor problem by starting with $\mathbf{J} = \sigma \mathbf{E}$. Finding the current flux through the same surface,

$$I = \int \mathbf{J} \cdot d\mathbf{A} = \sigma \int \mathbf{E} \cdot d\mathbf{A} = \frac{V}{R}$$

So

$$\frac{1}{V} \int \mathbf{E} \cdot d\mathbf{A} = \frac{1}{\sigma R} \qquad \text{(S.3.19.4)}$$

Equating (S.3.19.3) and (S.3.19.4), we find

$$\frac{1}{\sigma R} = \frac{4\pi C}{\varepsilon}$$

and finally

$$RC = \frac{\varepsilon}{4\pi\sigma} \qquad \text{(S.3.19.5)}$$

The parallel plate capacitor has the following capacitance and resistance:

$$C = \frac{\varepsilon L^2}{4\pi d} \qquad R = \frac{1}{\sigma} \frac{d}{L^2}$$

Thus, we confirm the general result for RC.

b) Find the potential at the surface of each conductor:

$$V_1 \approx \frac{q}{a} - \frac{q}{c} \qquad V_2 \approx -\frac{q}{b} + \frac{q}{c} \qquad \text{(S.3.19.6)}$$

So

$$C \approx \left(\frac{1}{a} + \frac{1}{b} - \frac{2}{c} \right)^{-1} \qquad \text{(S.3.19.7)}$$

c) Following the hint, consider the potential drop from each conductor to infinity when first $I_2 = 0$ and then $I_1 = 0$:

$$V_1 = -\int_a^\infty \mathbf{E} \cdot d\mathbf{l} = -\int_a^\infty \frac{\mathbf{J}}{\sigma} \cdot d\mathbf{l} = -\frac{I_1}{4\pi\sigma} \int_a^\infty \frac{dr}{r^2}$$

so

$$V_1 = \frac{I_1}{4\pi\sigma} \frac{1}{a} \qquad R_{11} = \frac{1}{4\pi\sigma a}$$

In the same fashion

$$R_{21} \approx \frac{1}{4\pi\sigma c}$$

$$R_{12} \approx \frac{1}{4\pi\sigma c}$$

$$R_{22} = \frac{1}{4\pi\sigma b}$$

Now, $V_1 = (R_{11} - R_{12}) I$ and $V_2 = (R_{21} - R_{22}) I$, where $I = I_1 = -I_2$. So we obtain for R

$$R = \frac{V_1 - V_2}{I} = R_{11} - R_{12} - R_{21} + R_{22} \approx \frac{1}{4\pi\sigma} \left(\frac{1}{a} + \frac{1}{b} - \frac{2}{c} \right) \quad \text{(S.3.19.8)}$$

d) Multiplying (S.3.19.7) by (S.3.19.8), we find the result in (S.3.19.5) with $\varepsilon = 1$

$$RC = \frac{1}{4\pi\sigma}$$

3.20 Charge Distribution in Inhomogeneous Medium (Boston)

From the Maxwell equation for $\mathbf{D} = \varepsilon\mathbf{E}$

$$\nabla \cdot \mathbf{D} = 4\pi\rho$$

we have

$$4\pi\rho = \nabla \cdot \mathbf{D} = \nabla \cdot (\varepsilon\mathbf{E}) = (\nabla\varepsilon) \cdot \mathbf{E} + \varepsilon\nabla \cdot \mathbf{E} \quad \text{(S.3.20.1)}$$

Substituting $\mathbf{J}/\sigma$ for $\mathbf{E}$ in (S.3.20.1) (for an isotropic medium where $\mathbf{J}$ is the current density), we find

$$4\pi\rho = (\nabla\varepsilon) \cdot \mathbf{E} + \varepsilon\nabla \cdot \left(\frac{\mathbf{J}}{\sigma} \right) = (\nabla\varepsilon) \cdot \mathbf{E} - \frac{\varepsilon}{\sigma^2}(\nabla\sigma) \cdot \mathbf{J} + \frac{\varepsilon}{\sigma}\nabla \cdot \mathbf{J}$$

From the condition that the current is stationary $\nabla \cdot \mathbf{J} = 0$ and substituting back for $\mathbf{J}$,

$$4\pi\rho = (\nabla\varepsilon) \cdot \mathbf{E} - \frac{\varepsilon}{\sigma}(\nabla\sigma) \cdot \mathbf{E} = -\frac{1}{\sigma}(\sigma\nabla\varepsilon - \varepsilon\nabla\sigma) \cdot \nabla\varphi$$

and so

$$\rho = -\frac{1}{4\pi\sigma}(\sigma\nabla\varepsilon - \varepsilon\nabla\sigma) \cdot \nabla\varphi$$

3.21   Green's Reciprocation Theorem (Stony Brook)

(See Problems 1.12 and 1.13 of Jackson, *Classical Electrodynamics*)

a) We may prove the theorem by considering the volume integral of the following expression:

$$\int_V \nabla\phi \cdot \nabla\phi' \, d^3x$$

Integrating by parts in two ways, we have

$$\int_V \nabla\phi \cdot \nabla\phi' \, d^3x = \int_S \phi\nabla\phi' \cdot d\mathbf{S} - \int_V \phi\nabla^2\phi' \, d^3x$$

$$= \int_S \phi'\nabla\phi \cdot d\mathbf{S} - \int_V \phi'\nabla^2\phi \, d^3x \qquad (S.3.21.1)$$

Now, $\nabla\phi \cdot \mathbf{n} = 4\pi\sigma$ (where $\mathbf{n}$ points opposite to the directed area of the surface $\mathbf{S}$) and $\nabla^2\phi = -4\pi\rho$, so dividing (S.3.21.1) by 4π yields the desired result:

$$\int_V \rho\phi' \, d^3x + \int_S \sigma\phi' \, dS = \int_V \rho'\phi \, d^3x + \int_S \sigma'\phi \, dS \qquad (S.3.21.2)$$

b) Let us introduce a second potential given by $\phi' = 2\pi\sigma' z$, corresponding to a surface charge density on the upper plate of σ' and on the lower plate of 0 (see Figure S.3.21). This introduced potential has no charge in the volume, and the real potential is zero on the plates so that the right-hand side of (S.3.21.2) is zero, yielding

$$\int_V q\delta(\mathbf{x} - \mathbf{z}_0) \, 2\pi\sigma' z \, d^3x + \int_{\text{upper plate}} \sigma \cdot 2\pi\sigma' l \, dS$$

$$= q \cdot 2\pi\sigma' z_0 + Q_i \cdot 2\pi\sigma' l = 0$$

where Q_i is the induced charge. So on the upper plate,

$$Q_i = -q\frac{z_0}{l}$$

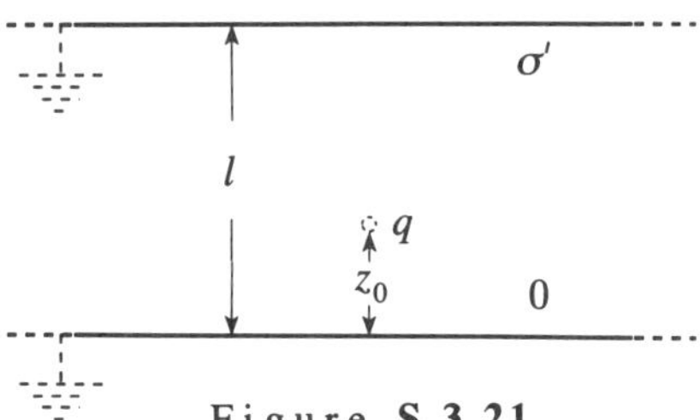

Figure S.3.21

3.22   Coaxial Cable and Surface Charge (Princeton)

Write down Laplace's equation in the region between the cylinders in cylindrical coordinates:

$$\nabla^2\phi = \frac{1}{\rho}\frac{\partial}{\partial\rho}\left(\rho\frac{\partial\phi}{\partial\rho}\right) + \frac{1}{\rho^2}\frac{\partial^2\phi}{\partial\varphi^2} + \frac{\partial^2\phi}{\partial z^2} = 0$$

By cylindrical symmetry, ϕ does not depend on φ (see Figure S.3.22), so

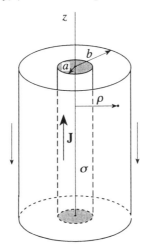

Figure S.3.22

Laplace's equation reduces to

$$\nabla^2\phi = \frac{1}{\rho}\frac{\partial}{\partial\rho}\left(\rho\frac{\partial\phi}{\partial\rho}\right) + \frac{\partial^2\phi}{\partial z^2} = 0 \qquad (S.3.22.1)$$

As usual, we look for a solution in the form

$$\phi(\rho, z) = R(\rho)Z(z)$$

and (S.3.22.1) becomes

$$\frac{1}{R}\frac{1}{\rho}\frac{\partial}{\partial\rho}\left(\rho\frac{\partial R}{\partial\rho}\right) + \frac{1}{Z}\frac{\partial^2 Z}{\partial z^2} = 0 \qquad (S.3.22.2)$$

From (S.3.22.2), we have

$$\frac{\partial^2 Z}{\partial z^2} - k^2 Z = 0 \qquad (S.3.22.3)$$

where k is a constant. By translational symmetry along the z-axis,

$$\frac{\partial E_z}{\partial z} = 0 = \frac{\partial^2 Z}{\partial z^2}$$

so $k = 0$, and

$$Z(z) = \alpha z + \beta$$

where α and β are constants. The radial part then becomes

$$\frac{\partial}{\partial \rho}\left(\rho\frac{\partial R}{\partial \rho}\right) = 0$$

$$\text{(S.3.22.4)}$$

$$R(\rho) = A\ln(\rho/\rho_0)$$

So

$$\phi(\rho, z) = (\alpha z + \beta)A\ln(\rho/\rho_0) \qquad \text{(S.3.22.5)}$$

Imposing the boundary condition that $\phi(\rho, 0) = 0$ leads to $\beta = 0$, and we have

$$\phi(\rho, z) = \tilde{A}z\ln(\rho/\rho_0) \qquad \text{(S.3.22.6)}$$

where $\tilde{A}$ is a constant ($\tilde{A} = \alpha A$). We may write the potential differences along the cable as

$$\phi(a, z) - \phi(a, 0) = \mathbf{E}\cdot\mathbf{z} = \frac{Jz}{\sigma}$$

$$\phi(b, z) - \phi(b, 0) = 0$$

$$\text{(S.3.22.7)}$$

where we used $\mathbf{J} = \sigma\mathbf{E}$, the uniformity of the current density $\mathbf{J}$ and isotropic electrical conductivity. We now have

$$\tilde{A}z\ln(a/\rho_0) = \frac{Jz}{\sigma}$$

$$\tilde{A}\ln(b/\rho_0) = 0$$

$$\text{(S.3.22.8)}$$

From (S.3.22.8), we find

$$\rho_0 = b \qquad \text{(S.3.22.9)}$$

$$\tilde{A} = \frac{J}{\sigma\ln(a/b)} \qquad \text{(S.3.22.10)}$$

For $\rho < a$, the potential is the same as on the surface, and from (S.3.22.7) and (S.3.22.9)

$$\phi(\rho, z) = \begin{cases} \dfrac{Jz\ln(\rho/b)}{\sigma\ln(a/b)} & a \leq \rho \leq b \\[3mm] \dfrac{Jz}{\sigma} & \rho \leq a \end{cases}$$

The surface charge density is

$$\sigma = \frac{E_\rho(a^+) - E_\rho(a^-)}{4\pi}$$

$$= \frac{1}{4\pi}\left(-\frac{\partial\phi_{a^+}}{\partial\rho}\bigg|_{\rho=a^+} + \frac{\partial\phi_{a^-}}{\partial\rho}\bigg|_{\rho=a^-}\right) = -\frac{Jz}{4\pi a\sigma\ln(a/b)}$$

where a^+ and a^- correspond to the points outside and inside the inner cylinder, respectively.

3.23 Potential of Charged Rod (Stony Brook)

a) The potential along the z-axis may be computed by integrating along the rod:

$$\phi(z) = \int_{-L/2}^{L/2} \frac{Q}{L}\frac{1}{z - \xi}\, d\xi = -\frac{Q}{L}\,\ln|\xi - z|\bigg|_{-L/2}^{L/2}$$

$$= -\frac{Q}{L}\ln\left|\frac{L/2 - z}{-L/2 - z}\right| = \frac{Q}{L}\ln\left|\frac{1 + u}{1 - u}\right| \qquad (S.3.23.1)$$

where $u \equiv L/2z$.

b) Since the problem has azimuthal symmetry, we may use the Legendre polynomials $P_l(\cos\theta)$ and rewrite the expansion in (P.3.23.1) (see §3.3 of Jackson, *Classical Electrodynamics* and Figure S.3.23)

$$\phi(r, \theta) = \sum_{l=0}^{\infty}\left(A_l r^l + \frac{B_l}{r^{l+1}}\right)P_l(\cos\theta)$$

Since we consider $r > L/2$, all the A_l are zero, and

$$\phi(r, \theta) = \sum_{l=0}^{\infty}\frac{B_l}{r^{l+1}}P_l(\cos\theta) \qquad (S.3.23.2)$$

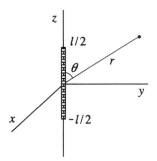

Figure S.3.23

We wish to equate (S.3.23.2) along the z-axis with (S.3.23.1) where $P_l(1) = 1$ for all l. We must rewrite $\ln |(1 + u)/(1 - u)|$ as a sum. Now,

$$\ln(1 + u) = \sum_{n=1}^{\infty} \frac{(-1)^{n+1} u^n}{n}$$

So

$$\ln \left| \frac{1 + u}{1 - u} \right| = \sum_{n=1}^{\infty} \left[(-1)^{n+1} - (-1)^{n+1}(-1)^n \right] \frac{u^n}{n} = 2 \sum_{n=1}^{\infty} \frac{u^{2n-1}}{n} \qquad \text{(S.3.23.3)}$$

Rewriting (S.3.23.1) using (S.3.23.3), we find

$$\phi(z) = \frac{2Q}{L} \sum_{n=1}^{\infty} \frac{1}{n} \left(\frac{L}{2} \right)^{2n-1} \frac{1}{z^{2n-1}} = Q \sum_{n=1}^{\infty} \frac{1}{n} \left(\frac{L}{2} \right)^{2n-2} \frac{1}{z^{2n-1}}$$

Replacing n by $l + 1$,

$$\phi(z) = Q \sum_{l=0}^{\infty} \frac{1}{l+1} \left(\frac{L}{2} \right)^{2l} \frac{1}{z^{2l+1}} \qquad \text{(S.3.23.4)}$$

Using (S.3.23.2) and (S.3.23.4), we have

$$\phi(r, \theta) = Q \sum_{l=0}^{\infty} \frac{1}{l+1} \left(\frac{L}{2} \right)^{2l} \frac{1}{r^{2l+1}} P_{2l}(\cos \theta)$$

3.24 Principle of Conformal Mapping (Boston)

A differentiable function $W(z) = U(x, y) + iV(x, y)$ satisfies the Cauchy–Riemann conditions

$$\frac{\partial U(x, y)}{\partial x} = \frac{\partial V(x, y)}{\partial y} \qquad \frac{\partial U(x, y)}{\partial y} = -\frac{\partial V(x, y)}{\partial x} \qquad \text{(S.3.24.1)}$$

To check that U and V satisfy Laplace's equation, differentiate (S.3.24.1)

$$\frac{\partial^2 U}{\partial x^2} = \frac{\partial}{\partial x}\left(\frac{\partial U}{\partial x}\right) = \frac{\partial}{\partial x}\left(\frac{\partial V}{\partial y}\right) = \frac{\partial}{\partial y}\left(\frac{\partial V}{\partial x}\right) = -\frac{\partial^2 U}{\partial y^2}$$

or

$$\frac{\partial^2 U(x,y)}{\partial x^2} + \frac{\partial^2 U(x,y)}{\partial y^2} = 0$$

Similarly,

$$\frac{\partial^2 V}{\partial x^2} = \frac{\partial}{\partial x}\left(-\frac{\partial U}{\partial y}\right) = -\frac{\partial}{\partial y}\left(\frac{\partial U}{\partial x}\right) = -\frac{\partial}{\partial y}\left(\frac{\partial V}{\partial y}\right) = -\frac{\partial^2 V}{\partial y^2}$$

and

$$\frac{\partial^2 V(x,y)}{\partial x^2} + \frac{\partial^2 V(x,y)}{\partial y^2} = 0$$

b) Orthogonality of the functions $\mathbf{F}$ and $\mathbf{G}$ also follows from (S.3.24.1):

$$\mathbf{F} \cdot \mathbf{G} = F_x\,G_x + F_y\,G_y = \frac{\partial U}{\partial x}\frac{\partial V}{\partial x} + \frac{\partial U}{\partial y}\frac{\partial V}{\partial y} = -\frac{\partial U}{\partial x}\frac{\partial U}{\partial y} + \frac{\partial U}{\partial y}\frac{\partial U}{\partial x} = 0$$

c) The electric field of an infinitely long charged wire passing through the origin is given by $\mathbf{E}_r = -(A/r)\hat{\mathbf{r}}$, where $A = -2\lambda$ and λ is the charge per unit length, r is the distance from the wire (see Figure S.3.24). The complex potential

$$W = A\ln z = A\ln\left(re^{i\varphi}\right) = A\ln r + Ai\varphi$$

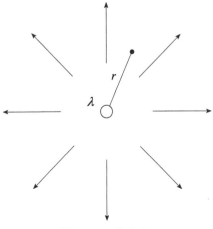

Figure S.3.24

So

$$U = A \ln \sqrt{x^2 + y^2}$$

$$V = A \tan^{-1}(y/x)$$

The fields $\mathbf{F}$ and $\mathbf{G}$ are given by

$$F_x = -\frac{\partial U}{\partial x} = -\frac{Ax}{r^2}$$

$$F_y = -\frac{\partial U}{\partial y} = -\frac{Ay}{r^2}$$

$$G_x = -\frac{\partial V}{\partial x} = \frac{Ay}{r^2}$$

$$G_y = -\frac{\partial V}{\partial y} = -\frac{Ax}{r^2}$$

Note how $\mathbf{F}$ and $\mathbf{G}$ satisfy the conditions of parts (a) and (b). The magnetic field of a similarly infinite line current can be described by the same potential.

3.25   Potential above Half Planes (Princeton)

This problem is symmetric for displacements along the z-axis, so we can consider this a two-dimensional problem in the x–y plane, a candidate for the method of conformal mapping. It can be seen that the function $w = \ln \eta$ ($\eta = x + iy$, $w = u + iv$) transforms the initial plane so that points at $y = 0$ for $x > 0$ map into the line $v = 0$, and the points at $y = 0$ for $x < 0$ map into the line $v = \pi$ (see Figure S.3.25). In the w plane

$$V = V_0 - \frac{2V_0}{\pi} v$$

so that at $v = 0$, $V = V_0$ and at $v = \pi$, $V = -V_0$. Again using

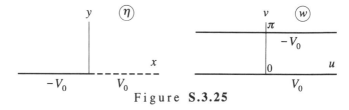

Figure **S.3.25**

$\eta = x + iy = \rho e^{i\varphi}$, we obtain

$$\ln(x + iy) = \ln\left(\rho e^{i\varphi}\right) = \ln\rho + i\varphi = u + iv$$

so that

$$\tan\varphi = \frac{y}{x}$$

We then have

$$v = \varphi = \tan^{-1}\frac{y}{x}$$

and

$$\phi(x, y) = V_0 - \frac{2V_0}{\pi}\tan^{-1}\frac{y}{x} = \frac{2V_0}{\pi}\left(\frac{\pi}{2} - \tan^{-1}\frac{y}{x}\right)$$

or using

$$\tan^{-1}\frac{y}{x} = \frac{\pi}{2} - \tan^{-1}\frac{x}{y}$$

we find

$$\phi(x, y) = \frac{2V_0}{\pi}\tan^{-1}\frac{x}{y}$$

We can check that $\phi(x, y)$ satisfies the boundary conditions.

3.26   Potential of Halved Cylinder (Boston, Princeton, Chicago)

This problem can be solved by several methods. We will use conformal mapping. Namely, we will try to find a function $w(z)$, where $z = x + iy$, to transform the curves of equal potential (in the cross section of the three-dimensional body) into parallel straight lines in the u–v plane, where $w = u + iv = f(x, y) + ig(x, y)$ with both f and g satisfying the Laplace's equation. We can easily find the solution for the potential problem in the w plane, and because of the properties of a conformal mapping (see Problem 3.24), the functions $u = f(x, y)$ or $v = g(x, y)$ will be a solution to the initial potential problem. For this problem, we can use the transformation (see, for instance, Spiegel, Schaum's Outline, *Complex Variables* or Kober, *Dictionary of Conformal Representations* for many useful conformal maps)

$$w = \ln\frac{a + z}{a - z} \tag{S.3.26.1}$$

This will transform a circle $R = a$ into two straight lines (see Figure S.3.26). The upper half of the cylinder will go into $v = \pi/2$, and the lower half will

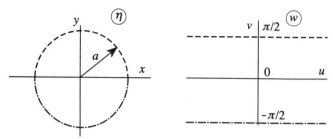

Figure S.3.26

go into $v = -\pi/2$. So we have

$$w = u + iv = \ln \frac{a + x + iy}{a - x - iy}$$

$$= \ln \frac{(a + x + iy)(a - x + iy)}{(a - x - iy)(a - x + iy)} = \ln \frac{a^2 - x^2 - y^2 + 2ayi}{(a - x)^2 + y^2}$$

$$(S.3.26.2)$$

Denote the argument of the natural log as $\rho e^{i\varphi}$, where ρ and φ are real. Then,

$$u + iv = \ln \rho e^{i\varphi} = \ln \rho + i\varphi$$

So $v = \varphi$. On the other hand,

$$\frac{a^2 - x^2 - y^2 + 2ayi}{(a - x)^2 + y^2} = \rho e^{i\varphi} \qquad (S.3.26.3)$$

For a complex number $m + in = \rho e^{i\varphi}$, we have

$$m^2 + n^2 = \rho \qquad (S.3.26.4)$$

$$\tan \varphi = \frac{n}{m} \qquad (S.3.26.5)$$

Using (S.3.26.5), we obtain from (S.3.26.3)

$$\tan \varphi = \frac{2ay}{a^2 - (x^2 + y^2)}$$

or

$$\varphi = v = \tan^{-1} \frac{2ay}{a^2 - (x^2 + y^2)}$$

So the potential which satisfies $\phi(v = -\pi/2) = -V_0$, $\phi(v = \pi/2) = V_0$ is

$$\phi = \frac{2V_0}{\pi} \tan^{-1} \frac{2ay}{a^2 - (x^2 + y^2)} \qquad (S.3.26.6)$$

On the z-axis

$$E_x = -\frac{\partial \phi}{\partial x}$$

$$= -\frac{2V_0}{\pi} \left\{ \frac{1}{\left(2ay/[a^2 - (x^2 + y^2)]\right)^2 + 1} \cdot \frac{2ay(2x)}{[a^2 - (x^2 + y^2)]^2} \right\} \Bigg|_{x=0,y=0} = 0$$

$$E_y = -\frac{\partial \phi}{\partial y}$$

$$= -\frac{2V_0}{\pi} \left\{ \frac{1}{\left(2ay/[a^2 - (x^2 + y^2)]\right)^2 + 1} \cdot \frac{2a\left[a^2 - (x^2 + y^2)\right] + 4ay^2}{[a^2 - (x^2 + y^2)]^2} \right\} \Bigg|_{x=0,y=0}$$

$$= -\frac{4V_0}{\pi a}$$

$$E_z = 0$$

A different solution to this problem may be found in Cronin, Greenberg, Telegdi, *University of Chicago Graduate Problems in Physics.*

3.27 Resistance of a Washer (MIT)

From the cylindrical symmetry of the washer, there is no radial dependence of the potential. We can therefore consider infinitesimal current rings flowing through the washer and sum them to obtain the total current and thereby the lumped resistance. Given that the potential difference between the faces of the cut washer is V, we have

$$dI = \frac{V a dr}{2\pi r \rho} \tag{S.3.27.1}$$

where r is the radius of one of the current rings. Integrating (S.3.27.1) from the inner to the outer radius,

$$I = \int_a^{2a} dI = \int_a^{2a} \frac{V a dr}{2\pi r \rho} = \frac{a \ln 2}{2\pi \rho} V$$

so the resistance

$$R = \frac{2\pi \rho}{a \ln 2} \approx 2.9 \frac{\pi \rho}{a}$$

The resistance of the washer is bounded by the resistances of a bar $a \times a$

in cross section and either $2\pi a$ or $4\pi a$ in length:

$$\frac{2\pi\rho}{a} = R_a < R < R_{2a} = \frac{4\pi\rho}{a}$$

3.28   Spherical Resistor (Michigan State)

The current density at point P may be written down immediately because of the cylindrical symmetry of the problem (see Figure S.3.28). The current

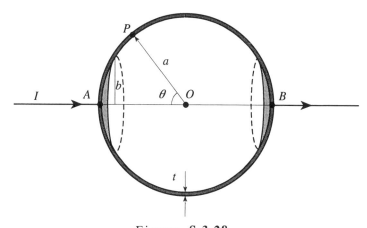

Figure S.3.28

I is divided evenly through 2π so that the current density J at each point in the spherical shell is

$$\mathbf{J}(\theta) = \frac{I}{2\pi a t \sin\theta}\mathbf{e}_\theta \qquad (S.3.28.1)$$

From the equation $\mathbf{J} = \sigma\mathbf{E}$, where σ is the conductivity of the shell, we obtain

$$V = \frac{1}{\sigma}\int_{\theta_1}^{\theta_2} \mathbf{J}\cdot d\mathbf{l}$$

where V is the potential difference between the two electrodes. So

$$R = \frac{V}{I} = \int_{\theta_1}^{\theta_2}\frac{a\,d\theta}{2\pi\sigma a t\sin\theta} = \frac{1}{2\pi\sigma t}\int_{\sin^{-1}(b/a)}^{\pi-\sin^{-1}(b/a)}\frac{d\theta}{\sin\theta} \qquad (S.3.28.2)$$

From the hint in the problem (which can by computed by using the substitution $\tan\theta/2 = t$), we can take the integral in (S.3.28.2):

$$R = -\frac{1}{2\pi\sigma t}\frac{1}{2}\ln\left[\frac{1+\cos\theta}{1-\cos\theta}\right]\Bigg|_{\sin^{-1}(b/a)}^{\pi-\sin^{-1}(b/a)}$$

$$= \frac{1}{2\pi\sigma t}\frac{1}{2}\ln\left[\frac{1+\sqrt{1-(b/a)^2}}{1-\sqrt{1-(b/a)^2}}\frac{1+\sqrt{1-(b/a)^2}}{1-\sqrt{1-(b/a)^2}}\right]$$

$$= \frac{1}{2\pi\sigma t}\ln\left[\frac{1+\sqrt{1-(b/a)^2}}{1-\sqrt{1-(b/a)^2}}\right]$$

As the radius b of the electrodes goes to zero, the resistance goes to infinity!

3.29 Infinite Resistor Ladder (Moscow Phys-Tech)

Define equivalent resistances R and R' as shown in Figure S.3.29a. By symmetry, the equivalent resistances attached to points A and C are both R. The total resistance between terminals A and C is the sum of the series resistance

$$R_{AC} = R + R = 2R \qquad (S.3.29.1)$$

Now, rewrite the ladder one rung back, as in Figure S.3.29b. The new resistance $R_{A'C'}$ should equal the original R_{AC}, given that an infinite ladder of resistors has a finite resistance between the terminals in the first place. Without calculating loops, it can be seen that R' does not contribute to the resistance $R_{A'C'}$ (Remove R'. What would be the potential difference between its terminals?). Ignoring B', we have the equivalent circuit in Figure S.3.29c. The loop on the right may be replaced by

$$R^* = \frac{2r \cdot 2R}{2r + 2R} = \frac{2rR}{r + R}$$

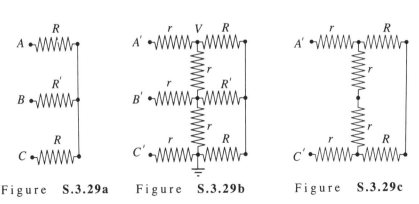

Figure S.3.29a   Figure S.3.29b   Figure S.3.29c

so

$$R_{A'C'} = r + R^* + r = 2r + \frac{2rR}{r+R} = 2\frac{r^2 + 2rR}{r+R} \qquad (S.3.29.2)$$

But, from (S.3.29.1),

$$R_{AC} = 2R = R_{A'C'} = 2\frac{r^2 + 2rR}{r+R}$$

We arrive at the quadratic equation

$$R^2 - rR - r^2 = 0$$

whose positive solution,

$$R = \frac{r + r\sqrt{5}}{2} = \frac{(\sqrt{5}+1)}{2}r$$

gives a resistance of $(\sqrt{5}+1)\,r$ for R_{AC}. Calculation of R_{AB} can be done in the same manner as for R_{AC}.

3.30   Semi-infinite Plate (Moscow Phys-Tech)

Consider the voltage difference between arbitrary points (E, F) a distance x_i from the end (see Figure S.3.30). We can write

$$V(x_i + dx) - V(x_i) = \alpha(x_i) V(x_i)\ dx \qquad (P.3.30.1)$$

Since the plate is semi-infinite, (P.3.30.1) is true for any other points (G, H) of the plate. So

$$\alpha(x_i) = \alpha(x_j) = \text{const} = \alpha \qquad (P.3.30.2)$$

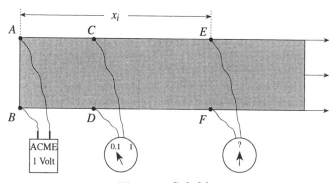

Figure S.3.30

Hence, for an arbitrary point a distance x from the end, we have

$$V(x + dx) - V(x) = \alpha V(x)\, dx \qquad \text{(P.3.30.3)}$$

or

$$\frac{dV}{dx} dx = \alpha V(x) dx \qquad \text{(P.3.30.4)}$$

$$V(x) = V_0 e^{\alpha x} \qquad \text{(P.3.30.5)}$$

From the condition $V(d) = V_0/10$, we have

$$\alpha = \frac{1}{d} \ln \frac{V_0/10}{V_0} = -\frac{1}{d} \ln 10 \qquad \text{(P.3.30.6)}$$

Therefore

$$V(x) = V_0 e^{-(x/d)\ln 10} = V_0 10^{-x/d}$$

For the values given in the problem, the voltage measured a distance x cm from the end gives the function

$$f(x) = 10^{-x}$$

3.31 Magnetic Field in Center of Cube (Moscow Phys-Tech)

The field can be calculated directly from the current flowing in each of the edges of the cube and then expressed in terms of B_0, but it is easier to use symmetry considerations. In Figure P.3.31a, the field is clearly parallel to the y-axis ($\hat{\mathbf{y}}$ is perpendicular to the face of the cube), so

$$\mathbf{B} = B_0 \hat{\mathbf{y}}$$

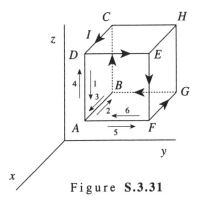

Figure S.3.31

For the problem at hand, we can add several of these "current loops" in order to produce the configuration given in Figure P.3.31b, noting that two overlapping but opposite current legs produce no field (see Figure S.3.31). From the figure, we see that there are now three faces of the cube with a current I flowing about their edges: $ABCD$, $AFGB$, and $ADEF$, producing the fields

$$\mathbf{B}_1 = B_0\hat{\mathbf{y}} \qquad \mathbf{B}_2 = B_0\hat{\mathbf{z}} \qquad \mathbf{B}_3 = -B_0\hat{\mathbf{x}}$$

So the total field $\mathbf{B} = B_0(-1, 1, 1)$, where the direction is along the diagonal of the cube and the magnitude is

$$|\mathbf{B}| = B_0\sqrt{3}$$

3.32 Magnetic Dipole and Permeable Medium (Princeton)

a) Use the method of images. We place another dipole $\mathbf{m}' = \alpha\mathbf{m}$ at the point O', at the same distance d on the other side of the plane separating the vacuum from the permeable medium (see Figure S.3.32). Compute the field in medium 1 as a superposition of the dipole fields from $\mathbf{m}$ and $\mathbf{m}'$. To find the field in medium 2, we put yet another dipole $\mathbf{m}'' = \beta\mathbf{m}$ at point O. We can write this in the form

$$\mathbf{B}^{(1)} = \mathbf{H}^{(1)} = \frac{3(\mathbf{m}\cdot\mathbf{n})\mathbf{n} - \mathbf{m}}{r^3} + \alpha\frac{3(\mathbf{m}\cdot\mathbf{n}')\mathbf{n}' - \mathbf{m}}{r'^3} \qquad \text{(S.3.32.1)}$$

$$\mathbf{H}^{(2)} = \beta\frac{3(\mathbf{m}\cdot\mathbf{n})\mathbf{n} - \mathbf{m}}{r^3} \qquad \text{(S.3.32.2)}$$

where $\mathbf{n}, \mathbf{n}'$ are unit vectors in the directions $\mathbf{r}, \mathbf{r}'$, respectively, and indices 1 and 2 correspond to media 1 and 2. As usual, we write the boundary

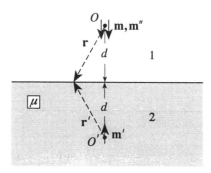

Figure S.3.32

conditions at some point P on the plane where $|\mathbf{r}| = |\mathbf{r}'|$

$$H_t^{(1)} = H_t^{(2)} \tag{S.3.32.3}$$

$$B_n^{(1)} = B_n^{(2)} = \mu H_n^{(2)} \tag{S.3.32.4}$$

Substituting (S.3.32.1) and (S.3.32.2) into (S.3.32.3) and (S.3.32.4), we obtain

$$\frac{3m\cos\theta\sin\theta}{r^3} + \alpha\frac{3m\cos\theta\sin\theta}{r^3} = \beta\frac{3m\cos\theta\sin\theta}{r^3} \tag{S.3.32.5}$$

$$m\frac{3\cos^2\theta - 1}{r^3} - \alpha m\frac{3\cos^2\theta - 1}{r^3} = \mu\beta m\frac{3\cos^2\theta - 1}{r^3} \tag{S.3.32.6}$$

From (S.3.32.5) and (S.3.32.6), we have

$$\begin{aligned} 1 + \alpha &= \beta \\ 1 - \alpha &= \mu\beta \end{aligned} \tag{S.3.32.7}$$

which yield

$$\beta = \frac{2}{\mu + 1}$$

$$\alpha = 1 - \mu\beta = -\frac{\mu - 1}{\mu + 1}$$

So the field in medium 2 is

$$\mathbf{H}^{(2)} = \frac{2}{\mu + 1}\frac{3(\mathbf{m}\cdot\mathbf{n})\mathbf{n} - \mathbf{m}}{r^3} \qquad \mathbf{B}^{(2)} = \mu\mathbf{H}^{(2)} = \frac{2\mu}{\mu + 1}\frac{3(\mathbf{m}\cdot\mathbf{n})\mathbf{n} - \mathbf{m}}{r^3}$$

b) The force acting on a dipole $\mathbf{m}$ is determined only by the field $\mathbf{B}^{(1)'}$ of the image dipole $\mathbf{m}'$ in medium 2

$$|\mathbf{F}| = F_x = \mathbf{m}\cdot\nabla B_x^{(1)'} = -3\frac{2\alpha m^2}{(2d)^4} = \frac{3}{8}\frac{\mu - 1}{\mu + 1}\frac{m^2}{d^4} \tag{S.3.32.8}$$

The dipole is attracted to the medium. The result for the equivalent problem of an electric dipole near a halfspace filled with an ideal conductor can be obtained from (S.3.32.8) by letting $\mu \to \infty$

$$|\mathbf{F}'| = F_x' = \frac{3}{8}\frac{p^2}{d^4}$$

where p is the electric dipole moment.

3.33 Magnetic Shielding (Princeton)

a) By analogy with electrostatics, we assume that the shell can be described by a magnetic dipole placed in the center of the shell for $r > a$ and try to satisfy boundary conditions for $\mathbf{H}$ and $\mathbf{B}$. We can write

$$\mathbf{H}_1 = \mathbf{B}_1 = \mathbf{H}_0 + \alpha \frac{3\left(\mathbf{H}_0 \cdot \mathbf{n}\right)\mathbf{n} - \mathbf{H}_0}{r^3} \qquad (\text{S.3.33.1})$$

$$\mathbf{H}_2 = \beta \mathbf{H}_0 + \gamma \frac{3\left(\mathbf{H}_0 \cdot \mathbf{n}\right)\mathbf{n} - \mathbf{H}_0}{r^3} \qquad (\text{S.3.33.2})$$

$$\mathbf{H}_3 = \delta \mathbf{H}_0 \qquad (\text{S.3.33.3})$$

where (S.3.33.1), (S.3.33.2), and (S.3.33.3) are written for areas 1, 2, and 3 outside the shell, at $b > r > a$, and inside the shell, respectively (see Figure S.3.33); α, β, γ, and δ are numerical factors that we shall find from

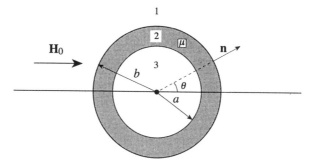

Figure S.3.33

the constitutive relation $\mathbf{B} = \mu \mathbf{H}$; and $\mathbf{n}$ is a unit vector parallel to $\mathbf{r}$. From (S.3.33.1)–(S.3.33.3), we can impose conditions for the normal components of $\mathbf{B}$ and the tangential components of $\mathbf{H}$, taken at the same angle θ.

$$\mu\beta H_0 \cos\theta + \frac{2\mu\gamma H_0}{b^3}\cos\theta = H_0\cos\theta + \frac{2\alpha H_0}{b^3}\cos\theta \qquad (\text{S.3.33.4})$$

$$\beta H_0 \sin\theta - \frac{\gamma H_0}{b^3}\sin\theta = H_0\sin\theta - \frac{\alpha H_0}{b^3}\sin\theta \qquad (\text{S.3.33.5})$$

$$\delta H_0 \cos\theta = \mu\beta H_0 \cos\theta + \frac{2\mu\gamma H_0}{a^3}\cos\theta \qquad (\text{S.3.33.6})$$

$$\delta H_0 \sin\theta = \beta H_0 \sin\theta - \frac{\gamma H_0}{a^3}\sin\theta \qquad (\text{S.3.33.7})$$

where (S.3.33.4) and (S.3.33.5) apply to interface 1–2, while (S.3.33.6) and (S.3.33.7) apply to interface 2–3. Dividing out the $H_0 \cos\theta$ and $H_0 \sin\theta$, appropriately, we obtain

$$\mu\beta + \frac{2\mu\gamma}{b^3} = 1 + \frac{2\alpha}{b^3} \qquad \text{(S.3.33.8)}$$

$$\beta - \frac{\gamma}{b^3} = 1 - \frac{\alpha}{b^3} \qquad \text{(S.3.33.9)}$$

$$\delta = \mu\beta + \frac{2\mu\gamma}{a^3} \qquad \text{(S.3.33.10)}$$

$$\delta = \beta - \frac{\gamma}{a^3} \qquad \text{(S.3.33.11)}$$

This system of four equations for the four numerical coefficients may be easily solved. Using (S.3.33.10) and (S.3.33.11), we find

$$\gamma = \frac{\beta(1-\mu)}{2\mu+1} a^3$$

and from (S.3.33.8) and (S.3.33.9), we obtain

$$\beta = \frac{3(2\mu+1)}{(2\mu+1)(\mu+2) - 2(a^3/b^3)(\mu-1)^2}$$

Now we can calculate δ, which is the attenuation factor we seek. Isolating δ from (S.3.33.10) and (S.3.33.11) and substituting β, we have

$$\delta = \frac{3\mu\beta}{1+2\mu} = \frac{9\mu}{(2\mu+1)(\mu+2) - 2(a^3/b^3)(\mu-1)^2}$$

b) In the limit of high permeability $\mu \gg 1$, we arrive at

$$\delta \approx \frac{9\mu}{2\mu^2 - 2(a^3/b^3)\mu^2} = \frac{9}{2\mu\,(1 - a^3/b^3)}$$

For $a = 0.5$ m, $b = 0.55$ m and $\mu = 10^5$

$$\delta \approx 2 \cdot 10^{-4}$$

3.34 Electromotive Force in Spiral (Moscow Phys-Tech)

The electromotive force $\mathcal{E}_e$ is given by

$$\mathcal{E}_e = -\frac{1}{c}\frac{d\Phi}{dt} = -\frac{1}{c}\frac{d}{dt}\int_S \mathbf{B}\cdot d\mathbf{S} = \frac{B_0\omega\sin\omega t}{2c}\int_0^{2\pi N} r^2(\varphi)\,d\varphi$$

$$= \frac{B_0\omega\sin\omega t}{2c}\int_0^{2\pi N}\left(\frac{h\varphi}{2\pi}\right)^2 d\varphi = \frac{\pi B_0\omega h^2 N^3 \sin\omega t}{3c} \tag{S.3.34.1}$$

where Φ is the magnetic flux, and we used $r(\varphi) = h\varphi/2\pi$. As another solution, in the approximation of a large number of coils, we can consider the nth coil to be a circle of radius $r_n = nh$. Then,

$$\mathcal{E}_e^{(n)} = -\frac{1}{c}\frac{d\Phi_n}{dt} = -\frac{1}{c}\frac{d}{dt}(BS_n) = -\frac{S_n}{c}\frac{dB}{dt} = \frac{S_n B_0\omega}{c}\sin\omega t$$

where $S_n = \pi r_n^2 = \pi h^2 n^2$ is the area of the nth coil. So we find for the electromotive force between the ends of a single coil

$$\mathcal{E}_e^{(n)} = \frac{\pi B_0\omega\sin\omega t}{c}r_n^2 = \frac{\pi B_0\omega h^2\sin\omega t}{c}n^2 \tag{S.3.34.2}$$

Summing up over all the coils from (S.3.34.2), we obtain the total electromotive force (S.3.34.1)

$$\mathcal{E}_e = \frac{\pi B_0\omega h^2\sin\omega t}{c}\sum_{n=0}^{N}n^2 \approx \frac{\pi B_0\omega h^2 N^3\sin\omega t}{3c} \tag{S.3.34.3}$$

where we have used the approximation

$$\sum_{n=0}^{N}n^2 = \frac{N(N+1)(2N+1)}{6} \approx \frac{N^3}{3}$$

for $N \gg 1$.

3.35 Sliding Copper Rod (Stony Brook, Moscow Phys-Tech)

a) In this problem, $\mathbf{B} = B_0\hat{\mathbf{z}}$, so the magnetic flux through the surface limited by the rod and the rails changes as a result of the change of the surface area S (since the rod is moving). This gives rise to an electromotive force $\mathcal{E}_e$:

$$\mathcal{E}_e = -\frac{1}{c}\frac{d\Phi}{dt}$$

where $\Phi = \int_s \mathbf{B} \cdot d\mathbf{S}$ is the magnetic flux, so

$$\mathcal{E}_e = -\frac{1}{c}\frac{d(B_0 S)}{dt} = -\frac{1}{c}B_0 l\frac{dy(t)}{dt} = -\frac{1}{c}B_0 lv$$

In its turn, $\mathcal{E}_e$ produces the current through the rod:

$$I = \frac{\mathcal{E}_e}{R} = \frac{\mathcal{E}_e}{l/\sigma A} = -\frac{1}{c}vB_0\sigma A$$

where R is the resistance of the rod, A is its cross section, and σ is the conductivity of the rod. The force acting on the rod is

$$\mathbf{F} = \frac{I}{c}lB_0\left(\hat{\mathbf{x}} \times \hat{\mathbf{z}}\right) = -\frac{1}{c^2}vB_0^2 lA\sigma\hat{\mathbf{y}}$$

On the other hand, $\mathbf{F} = m\dot{\mathbf{v}} = m\dot{v}\hat{\mathbf{y}} = \rho_m Al\dot{v}\hat{\mathbf{y}}$. So we have

$$\rho_m Al\dot{v} = -\frac{v}{c^2}B_0^2 lA\sigma \qquad \dot{v} = -\frac{B_0^2\sigma}{\rho_m c^2}v$$

$$v = v_0 e^{-t/\tau} \qquad \tau = \frac{\rho_m c^2}{B_0^2\sigma}$$

b) For an estimate we can take that the rod practically stopped when $t/\tau = 10$. (It is good enough for an estimate, since for $t/\tau = 10$ the final velocity is $v_f = 5 \cdot 10^{-5}v_0$ and for $t/\tau = 20$, $v_f = 2 \cdot 10^{-9}v_0$.) So for $t/\tau = 10$,

$$t = 10\tau = 10\frac{\rho_m c^2}{B_0^2\sigma} := 10\frac{8.9 \cdot (3 \cdot 10^{10})^2}{1 \cdot 5 \cdot 10^{17}}$$

$$\approx 1.6 \cdot 10^5 \sec \approx 44\,\text{hours}$$

c) The kinetic energy of the rod is $T = Mv^2/2 = (Mv_0^2/2)e^{-2t/\tau}$, where M is the total mass of the rod. We can simply take the derivative of this kinetic energy per unit volume:

$$\frac{1}{V}\frac{dT}{dt} = -\frac{Mv_0^2}{V\tau}e^{-2t/\tau} = -\frac{\rho_m v_0^2}{\tau}e^{-2t/\tau}$$

where V is the volume of the rod. On the other hand, the Joule heating per unit volume

$$w = \frac{W}{V} = \frac{1}{V}\frac{\mathcal{E}_e^2}{R} = \frac{1}{V}\frac{\mathcal{E}_e^2}{l/\sigma A} = \frac{1}{c^2}\frac{v^2 B_0^2 l^2}{Vl}\sigma A = \frac{\rho_m}{\tau}v_0^2 e^{-2t/\tau}$$

so

$$w = -\frac{1}{V}\frac{dT}{dt}$$

3.36   Loop in Magnetic Field (Moscow Phys-Tech, MIT)

The magnetic force acting on the loop is proportional to its magnetic moment, which is proportional to the current flowing through the loop. The current I, in turn, is proportional to the rate of change of the magnetic flux through the loop, since $I = \mathcal{E}_e/R$, where $\mathcal{E}_e$ is the electromotive force and R is the resistance of the loop. We have

$$\mathcal{E}_e = -\frac{1}{c}\frac{d\Phi}{dt} \tag{S.3.36.1}$$

The magnetic flux Φ in (S.3.36.1) is given by

$$\Phi = BS = B_0(1 + \kappa z)S \tag{S.3.36.2}$$

where S is the area of the loop. From (S.3.36.1),

$$\mathcal{E}_e = -\frac{1}{c}\frac{d\Phi}{dt} = -\frac{1}{c}B_0\kappa\frac{dz}{dt}S \tag{S.3.36.3}$$

But dz/dt is the velocity of the loop. So the electromotive force increases with the velocity, and therefore the magnetic force F_m acting on the loop also increases with velocity, while the only other force, gravity, acting in the opposite direction, is constant. Therefore, the velocity will increase until $F_m = mg$. From energy conservation, the work done by gravity during this stationary motion goes into the Joule heating of the loop:

$$mg\Delta z = I^2 R\Delta t \tag{S.3.36.4}$$

But, since the velocity is constant,

$$\frac{\Delta z}{\Delta t} = v = \frac{I^2 R}{mg} = \frac{\mathcal{E}_e^2}{Rmg} = \frac{B_0^2}{c^2}\frac{\kappa^2 v^2 S^2}{Rmg} \tag{S.3.36.5}$$

where we substituted $\mathcal{E}_e$ from (S.3.36.3) again using $v = dz/dt$. From (S.3.36.5), we can find

$$v = \frac{c^2 Rmg}{B_0^2\kappa^2 S^2} \tag{S.3.36.6}$$

Now, substituting

$$R = \rho\frac{\pi D}{\pi d^2/4} = \rho\frac{4D}{d^2} \tag{S.3.36.7}$$

and

$$m = \rho_m V = \rho_m\frac{\pi d^2}{4}\pi D \tag{S.3.36.8}$$

into (S.3.36.6), we obtain

$$v = \frac{c^2 \rho \rho_m g \pi^2 D^2}{B_0^2 \kappa^2 (\pi D^2/4)^2} = \frac{16 c^2 \rho \rho_m g}{B_0^2 \kappa^2 D^2}$$

3.37 Conducting Sphere in Constant Magnetic Field (Boston)

In the frame K' moving with velocity $\mathbf{v}$, we have, to lowest order in v/c (see Problem 3.51),

$$\mathbf{E}' = \mathbf{E} + \frac{1}{c}\mathbf{v} \times \mathbf{B}$$

$$\mathbf{B}' = \mathbf{B} - \frac{1}{c}\mathbf{v} \times \mathbf{E}$$
(S.3.37.1)

In the lab frame K, we have $E = 0$ and $\mathbf{B} = B_0 \hat{\mathbf{y}}$. Using (S.3.37.1) for the frame K', we have an electric field in this frame

$$\mathbf{E}' = \frac{1}{c}\mathbf{v} \times \mathbf{B} = \frac{1}{c} v B_0 (\hat{\mathbf{x}} \times \hat{\mathbf{y}}) = \frac{1}{c} v B_0 \hat{\mathbf{z}} = E_0' \hat{\mathbf{z}} \qquad \text{(S.3.37.2)}$$

where $E_0' = v B_0/c$. Now, we have a perfectly conducting sphere in a constant electric field, so we may write the potential outside the sphere in the form (see Problem 3.4)

$$\phi = -E_0' r \cos\theta \left(1 - \frac{R^3}{r^3}\right) \qquad \text{(S.3.37.3)}$$

where θ is the angle between $\mathbf{r}$ and $\mathbf{E}'$, or, in this case, between $\mathbf{r}$ and $\hat{\mathbf{z}}$ (the origin of the spherical coordinates is at the center of the sphere). The surface charge density σ is given by

$$\sigma = \frac{1}{4\pi} E_n = -\frac{1}{4\pi} \left.\frac{\partial \phi}{\partial r}\right|_{r=R} = -\frac{1}{4\pi}(-3 E_0' \cos\theta) = \frac{3}{4\pi} E_0' \cos\theta$$

Finally, substituting E_0' from (S.3.37.2), we have

$$\sigma = \frac{3}{4\pi} \frac{v}{c} B_0 \cos\theta \qquad \text{(S.3.37.4)}$$

3.38   Mutual Inductance of Line and Circle (Michigan)

Label the circular wire conductor 1 and the straight wire conductor 2 (see Figure S.3.38). Recall that the mutual inductance of two conductors is given by

$$M_{12} = \frac{\Phi_{12}}{cI_2} \qquad (S.3.38.1)$$

where I_2 is the current flowing in conductor 2 and Φ_{12} is the magnetic flux from conductor 2 through the surface bounded by conductor 1 (In modi-

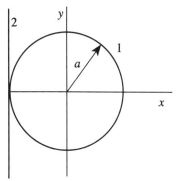

Figure S.3.38

fied Gaussian units, $M_{12} = c\Phi_{12}/I_2$. See, for instance, Jackson, *Classical Electrodynamics*, p.820).

$$\Phi_{12} = \int_S \mathbf{B}_2 \cdot d\mathbf{S} \qquad (S.3.38.2)$$

where $\mathbf{B}_2$ is the magnetic field produced by the straight wire and $d\mathbf{S}$ is the element of area of the loop. The magnitude of the field due to a current in an infinite wire

$$B_2 = \frac{2I_2}{rc} \qquad (S.3.38.3)$$

is perpendicular to the plane of the loop. So, from (S.3.38.1)–(S.3.38.3), we have

$$M_{12} = \frac{2}{c^2} \int_S \frac{dS}{r} \qquad (S.3.38.4)$$

Let $x = 0$, $y = 0$ be the center of the circle, and locate the infinite wire at $x = -a$. Then (S.3.38.4) becomes

$$M_{12} = \frac{2}{c^2} \int_{-a}^{a} \frac{2y\,dx}{a+x} = \frac{4}{c^2} \int_{-a}^{a} \frac{\sqrt{a^2-x^2}}{a+x}\,dx = \frac{4a}{c^2} \int_{-1}^{1} \sqrt{\frac{1-u}{1+u}}\,du \ (S.3.38.5)$$

Using the substitution $u = \cos 2\theta$, we obtain

$$M_{12} = -\frac{4a}{c^2} \int_{\pi/2}^{0} \frac{\sin\theta}{\cos\theta} 4\sin\theta\cos\theta \, d\theta = \frac{16a}{c^2} \int_{0}^{\pi/2} \sin^2\theta \, d\theta = \frac{4\pi a}{c^2}$$

3.39 Faraday's Homopolar Generator (Stony Brook, Michigan)

a) Consider an electron at a distance r from the axle (see Figure S.3.39). It experiences a Lorentz force

$$\mathbf{F} = \frac{e}{c}\left(\mathbf{v} \times \mathbf{B}\right) \tag{S.3.39.1}$$

with $\mathbf{v} = \boldsymbol{\omega} \times \mathbf{r}$, so we have a radial force $\mathbf{F}_r$ acting on the electron:

$$\mathbf{F}_r = \frac{e}{c}\left[(\boldsymbol{\omega} \times \mathbf{r}) \times \mathbf{B}\right] = \frac{e}{c}\omega B \mathbf{r} \tag{S.3.39.2}$$

where e is the electron charge. Therefore, the equivalent electric field $E = -(1/c)\omega Br$, and the voltage between C_2 and C_1 is

$$V = -\int_{0}^{r_0} \mathbf{E} \cdot d\mathbf{r} = \frac{1}{c}\omega B \int_{0}^{r_0} r \, dr = \frac{\omega B r_0^2}{2c}$$

The current i through the resistor R is given by

$$i = \frac{V}{R} = \frac{\omega B r_0^2}{2Rc} \tag{S.3.39.3}$$

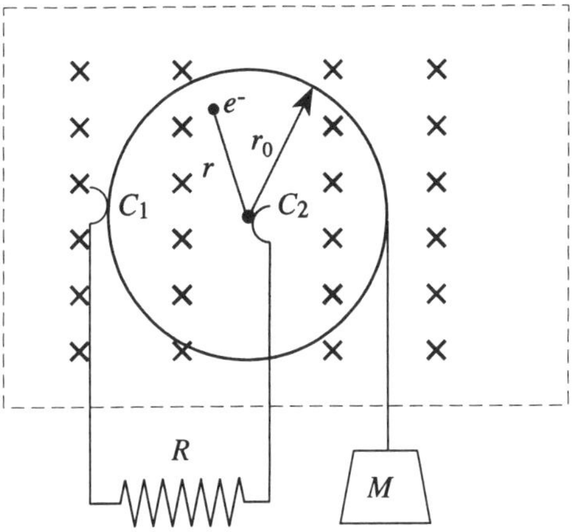

Figure **S.3.39**

b) The power P dissipated in the resistance can be found from (S.3.39.3)

$$P = i^2 R = \frac{\omega^2 B^2 r_0^4}{4Rc^2} = \frac{\dot{\varphi}^2 B^2 r_0^4}{4Rc^2}$$

The kinetic energy of the disk

$$T = \frac{I\omega^2}{2} = \frac{I\dot{\varphi}^2}{2} \tag{S.3.39.4}$$

where I is the moment of inertia of the disk. From energy conservation, we may write

$$\frac{d}{dt}\left(\frac{I\dot{\varphi}^2}{2}\right) + \frac{B^2 r_0^4}{4Rc^2}\dot{\varphi}^2 = Mgr_0\dot{\varphi} \tag{S.3.39.5}$$

For a constant angular velocity $\dot{\varphi} = \omega_f$, we have

$$\frac{B^2 r_0^4}{4Rc^2}\omega_f^2 = Mgr_0\omega_f \tag{S.3.39.6}$$

So

$$\omega_f = \frac{4MgRc^2}{B^2 r_0^3} \tag{S.3.39.7}$$

and

$$i_f = \frac{\omega_f B r_0^2}{2Rc} = \frac{2Mgc}{Br_0} \tag{S.3.39.8}$$

3.40 Current in Wire and Poynting Vector (Stony Brook, MIT)

a) Let us calculate the flux of the Poynting vector. Introduce cylindrical coordinates with unit vectors $\mathbf{e}_\rho$, $\mathbf{e}_\theta$, and $\hat{\mathbf{z}}$. Current flows along the wire in the z direction and the electric field $\mathbf{E} = E\hat{\mathbf{z}}$. Using one of Maxwell's equations in vacuum, the fact that conditions are stationary, and Stokes' theorem,

$$\nabla \times \mathbf{B} = \frac{4\pi}{c}\mathbf{J} + \frac{1}{c}\frac{\partial \mathbf{E}}{\partial t} \tag{S.3.40.1}$$

$$\int_A \nabla \times \mathbf{B} \cdot d\mathbf{A} = \oint_C \mathbf{B} \cdot d\mathbf{l} = \frac{4\pi}{c}\int_A \mathbf{J} \cdot d\mathbf{A} \tag{S.3.40.2}$$

where J is the current density and A is the surface. At any given radius

r, B_θ is constant, so we have

$$2\pi r B = \frac{4\pi}{c} J\pi r^2 \qquad \mathbf{B} = \frac{2\pi}{c} Jr \mathbf{e}_\theta \qquad \text{(S.3.40.3)}$$

$$\mathbf{S} = \frac{c}{4\pi} \mathbf{E} \times \mathbf{B} = \frac{c}{4\pi} \frac{2\pi}{c} JrE \left(\hat{\mathbf{z}} \times \mathbf{e}_\theta \right) = -\frac{1}{2} JrE \mathbf{e}_\rho \qquad \text{(S.3.40.4)}$$

Using the relation between current density and total current $J = I/(\pi b^2)$:

$$\mathbf{S} = -\frac{IEr}{2\pi b^2} \mathbf{e}_\rho \qquad \mathbf{S}(b) = -\frac{IE}{2\pi b} \mathbf{e}_\rho$$

b) The Poynting flux per unit length is then $\mathbf{S} \cdot 2\pi \mathbf{b} = -IE$. So the flux enters the wire, and we see that the dissipated power per unit length IE is equal to the total incoming S-flux, in agreement with Poynting's theorem:

$$\frac{\partial u}{\partial t} = -\mathbf{J} \cdot \mathbf{E} - \nabla \cdot \mathbf{S} \qquad \text{(S.3.40.5)}$$

where u is the energy density. Under stationary conditions such as ours

$$\frac{\partial u}{\partial t} = 0$$

and we have

$$\int_V \mathbf{J} \cdot \mathbf{E} \, d^3x = -\int_V \nabla \cdot \mathbf{S} \, dV = -\int_A \mathbf{S} \cdot d\mathbf{A} = IE$$

3.41   Box and Impulsive Magnetic Field (Boston)

a) From the Maxwell equation

$$\nabla \times \mathbf{E} = -\frac{1}{c} \frac{\partial \mathbf{B}}{\partial t} \qquad \text{(S.3.41.1)}$$

we can find the electric field induced in the box. We have for the $\hat{\mathbf{x}}$, $\hat{\mathbf{y}}$, and $\hat{\mathbf{z}}$ components

$$\hat{\mathbf{x}} \left(\frac{\partial E_z}{\partial y} - \frac{\partial E_y}{\partial z} \right) = 0$$

$$\hat{\mathbf{z}} \left(\frac{\partial E_y}{\partial x} - \frac{\partial E_x}{\partial y} \right) = 0 \qquad \text{(S.3.41.2)}$$

$$-\hat{\mathbf{y}} \left(\frac{\partial E_z}{\partial x} - \frac{\partial E_x}{\partial z} \right) = -\frac{1}{c} \frac{\partial B}{\partial t} \hat{\mathbf{y}}$$

From these equations, we obtain

$$E_y = E_z = 0$$

$$E_x = -\frac{1}{c}\frac{\partial B}{\partial t}z + E_0$$

where E_0 is some constant. From the equation for the force

$$\frac{d\mathbf{P}}{dt} = \left(-\sigma ab E_x \Big|_{z=0} + \sigma ab E_x \Big|_{z=h} \right) \hat{\mathbf{x}} = \sigma abh \left(-\frac{1}{c}\frac{\partial B}{\partial t} \right) \hat{\mathbf{x}}$$

we obtain the impulse received by the box when the magnetic field goes to zero

$$\mathbf{P} = \sigma abh \int\limits_0^t -\frac{1}{c}\frac{\partial B}{\partial t'}\, dt'\,\hat{\mathbf{x}} = \frac{1}{c}\sigma abh B_0 \hat{\mathbf{x}} \qquad (\text{S.3.41.3})$$

b) The initial momentum may be found from the Poynting vector

$$\mathbf{S} = \frac{c}{4\pi}\left[\mathbf{E}\times\mathbf{H}\right] = \frac{c}{4\pi}\begin{vmatrix} \hat{\mathbf{x}} & \hat{\mathbf{y}} & \hat{\mathbf{z}} \\ 0 & 0 & -4\pi\sigma \\ 0 & B_0 & 0 \end{vmatrix} = c\sigma B_0 \hat{\mathbf{x}} \qquad (\text{S.3.41.4})$$

$$\mathbf{P} = \int_V \frac{\mathbf{S}}{c^2}\, dV = \frac{1}{c}\sigma abh B_0 \hat{\mathbf{x}} \qquad (\text{S.3.41.5})$$

which is the same result as (S.3.41.3)

3.42 Coaxial Cable and Poynting Vector (Rutgers)

As in Problem 3.22, we have Laplace's equation in cylindrical coordinates whose solution is

$$\phi(\rho, z) = (\alpha z + \beta)\ln\left(\frac{\rho}{\rho_0}\right) \qquad (\text{S.3.42.1})$$

From the boundary conditions,

$$\phi(\rho, 0) = 0 \qquad \beta = 0$$

$$\phi(\rho, z) = \alpha z \ln\left(\frac{\rho}{\rho_0}\right) \qquad (\text{S.3.42.2})$$

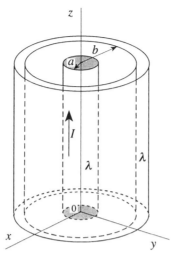

Figure S.3.42

Integrating the voltage drop along the cable (see Figure S.3.42), we find

$$-I\lambda z = \phi(a, z) - \phi(a, 0)$$
$$I\lambda z = \phi(b, z) - \phi(b, 0)$$

(S.3.42.3)

and so

$$\alpha = -\frac{2I\lambda}{\ln(a/b)} \qquad \rho_0 = \sqrt{ab}$$

resulting in

$$\phi(\rho, z) = -I\lambda z \frac{\ln[\rho^2/(ab)]}{\ln(a/b)} = \begin{cases} -I\lambda z, & \rho = a \\ \\ I\lambda z, & \rho = b \end{cases}$$

(S.3.42.4)

The electric field

$$\mathbf{E} = -\nabla \phi = -\frac{\partial \phi}{\partial \rho}\mathbf{e}_\rho - \frac{\partial \phi}{\partial z}\hat{\mathbf{z}} = \frac{2I\lambda z}{\rho \ln(a/b)}\mathbf{e}_\rho + \frac{I\lambda \ln[\rho^2/(ab)]}{\ln(a/b)}\hat{\mathbf{z}} \quad \text{(S.3.42.5)}$$

b) The magnetic field in the region $a < \rho < b$ can be found from

$$\oint \mathbf{H}_\varphi \cdot d\mathbf{l} = \frac{4\pi}{c}I$$

$$2\pi\rho H_\varphi = \frac{4\pi}{c}I$$

$$\mathbf{H} = H_\varphi \mathbf{e}_\varphi = \frac{2I}{c\rho}\mathbf{e}_\varphi$$

(S.3.42.6)

c) The Poynting vector is

$$\mathbf{S} = \frac{c}{4\pi}(\mathbf{E} \times \mathbf{H}) \tag{S.3.42.7}$$

Transforming $\mathbf{E}$ into Cartesian coordinates (see Appendix), we have

$$E_x = E_\rho \cos\varphi - E_\varphi \sin\varphi$$
$$E_y = E_\rho \sin\varphi + E_\varphi \cos\varphi \tag{S.3.42.8}$$
$$E_z = E_z$$

The same transformation applies to $\mathbf{H}$, so we obtain

$$\mathbf{E} \times \mathbf{H} = \begin{vmatrix} \hat{\mathbf{x}} & \hat{\mathbf{y}} & \hat{\mathbf{z}} \\ \\ E_\rho \cos\varphi & E_\rho \sin\varphi & E_z \\ \\ -H_\varphi \sin\varphi & H_\varphi \cos\varphi & 0 \end{vmatrix} \tag{S.3.42.9}$$

$$= -E_z H_\varphi \cos\varphi\, \hat{\mathbf{x}} - E_z H_\varphi \sin\varphi\, \hat{\mathbf{y}} + E_\rho H_\varphi\, \hat{\mathbf{z}}$$

$$= -E_z H_\varphi\, \mathbf{e}_\rho + E_\rho H_\varphi \hat{\mathbf{z}}$$

So

$$\mathbf{S} = -\frac{c}{4\pi}\frac{I\lambda \ln[\rho^2/(ab)]}{\ln(a/b)}\frac{2I}{c\rho}\mathbf{e}_\rho + \frac{c}{4\pi}\frac{2I\lambda z}{\rho \ln(a/b)}\frac{2I}{c\rho}\hat{\mathbf{z}}$$

$$= -\frac{I^2\lambda \ln[\rho^2/(ab)]}{2\pi\rho \ln(a/b)]}\mathbf{e}_\rho + \frac{I^2\lambda z}{\pi\rho^2 \ln(a/b)}\hat{\mathbf{z}} \tag{S.3.42.10}$$

We now write the flux F^{i} and F^{o} into the inner and outer conductors, respectively, from (S.3.42.10):

$$F^{\mathrm{i}} = \int_A \mathbf{S} \cdot d\mathbf{A} = \frac{I^2\lambda \ln(a/b)}{2\pi a \ln(a/b)}2\pi a \cdot l = I^2\lambda l = I^2 R \tag{S.3.42.11}$$

$$F^{\mathrm{o}} = \int_A \mathbf{S} \cdot d\mathbf{A} = \frac{I^2\lambda \ln(a/b)}{2\pi b \ln(a/b)}2\pi b \cdot l = I^2\lambda l = I^2 R \tag{S.3.42.12}$$

where R is the resistance of a length l of each conductor. The total flux going into the conductors $F = F^{\mathrm{i}} + F^{\mathrm{o}} = 2I^2 R$, which corresponds to the

Joule heating of the conductors (see also Problem 3.40). Since there is no current in the vacuum between the conductors and the conditions are stationary, Poynting's theorem (see (S.3.40.5)) gives

$$\nabla \cdot \mathbf{S} = 0 \qquad (S.3.42.13)$$

The total flux is zero. There must also be a corresponding negative flux F^e into the volume through the ends to satisfy Poynting's theorem. Indeed

$$F^e = \int_A S_z \hat{\mathbf{z}} \cdot d\mathbf{A} = \left(\frac{l}{2} - \frac{-l}{2}\right) \int_a^b \frac{I^2 \lambda}{\pi \rho^2 \ln(a/b)} \, 2\pi\rho \, d\rho$$

$$= \frac{2\pi I^2 \lambda l \ln(b/a)}{\pi \ln(a/b)} = -2I^2 \lambda l = -2I^2 R \qquad (S.3.42.14)$$

as expected.

3.43 Angular Momentum of Electromagnetic Field (Princeton)

By analogy with the electric dipole, we can write the magnetic field from the magnetic dipole $\mathbf{M}$ as

$$\mathbf{H} = -\frac{\mathbf{M}}{r^3} + \frac{3(\mathbf{M} \cdot \mathbf{r})\mathbf{r}}{r^5} \qquad (S.3.43.1)$$

where r is the distance from the center of the spheres (see Figure S.3.43). The electric field is nonzero only between the spheres:

$$E_{a<r<b} = \frac{q\mathbf{r}}{r^3} \qquad (S.3.43.2)$$

The electromagnetic angular momentum is given by the volume integral of $\mathbf{r} \times \mathbf{g}$, where $\mathbf{g}$ is the electromagnetic momentum density (see, for in-

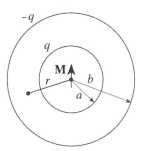

Figure S.3.43

stance, Jackson, *Classical Electrodynamics*, §6.13). The electromagnetic momentum density is

$$\mathbf{g} = \frac{\mathbf{S}}{c^2} \tag{S.3.43.3}$$

where $\mathbf{S}$ is the Poynting vector and c is the speed of light. Using the definition of the Poynting vector

$$\mathbf{S} = \frac{c}{4\pi} \left[\mathbf{E} \times \mathbf{H} \right] \tag{S.3.43.4}$$

we obtain for our field configuration

$$\mathbf{L}_{\text{em}} = \int_V \left(\mathbf{r} \times \mathbf{g} \right) dV = \frac{1}{4\pi c} \int_V \left[\mathbf{r} \times \left(\mathbf{E} \times \mathbf{H} \right) \right] dV$$

$$= \frac{q}{4\pi c} \int_V \frac{1}{r^3} \mathbf{r} \times \left[\mathbf{r} \times \left(-\frac{\mathbf{M}}{r^3} + \frac{3(\mathbf{M} \cdot \mathbf{r})\mathbf{r}}{r^5} \right) \right] dV$$

$$= -\frac{q}{4\pi c} \int_V \frac{1}{r^3} \left[\mathbf{r} \times \left(\mathbf{r} \times \frac{\mathbf{M}}{r^3} \right) \right] dV$$

$$= \frac{q}{4\pi c} \left[\int_V \frac{r^2}{r^3} \frac{\mathbf{M}}{r^3} dV - \int_V \frac{\mathbf{r}}{r^3} \frac{(\mathbf{r} \cdot \mathbf{M})}{r^3} dV \right] \tag{S.3.43.5}$$

(S.3.43.5) may be rewritten in the form

$$\mathbf{L}_{\text{em}} = \frac{q\mathbf{M}}{4\pi c} \int_a^b \frac{4\pi r^2 dr}{r^4} - \frac{q}{4\pi c} \int_V \frac{(x\hat{\mathbf{x}} + y\hat{\mathbf{y}} + z\hat{\mathbf{z}}) \, \mathbf{r} \cdot \mathbf{M}}{r^6} dV \tag{S.3.43.6}$$

Choose spherical coordinates with the z-axis in the direction of the vector $\mathbf{M}$, taking into account that

$$\mathbf{r} = (x, y, z) = (r \sin\theta \cos\varphi, r \sin\theta \sin\varphi, r \cos\theta)$$

We notice that the $\hat{\mathbf{x}}$ and $\hat{\mathbf{y}}$ components of the second integral vanish, and

$$\mathbf{L}_{\text{em}} = \frac{q\mathbf{M}}{c} \int_a^b \frac{dr}{r^2} - \frac{q\hat{\mathbf{z}}}{4\pi c} \int_0^{2\pi} d\varphi \int_a^b \int_0^\pi \frac{r\cos\theta \, rM\cos\theta}{r^4} \, dr \sin\theta \, d\theta$$

$$= \frac{q\mathbf{M}}{c} \left(\frac{1}{a} - \frac{1}{b} \right) - \frac{q\mathbf{M}}{2c} \left(\frac{1}{a} - \frac{1}{b} \right) \int_0^\pi \cos^2\theta \sin\theta \, d\theta$$

$$= \frac{q\mathbf{M}}{c} \left(\frac{1}{a} - \frac{1}{b} \right) + \frac{q\mathbf{M}}{2c} \left(\frac{1}{a} - \frac{1}{b} \right) (-2/3) = \frac{2}{3} \frac{q\mathbf{M}}{c} \left(\frac{1}{a} - \frac{1}{b} \right)$$

3.44   Plane Wave in Dielectric (Stony Brook, Michigan)

a) We assume that the dielectric constant is essentially real (no dissipation). For a monochromatic wave travelling in the z direction with $\mathbf{E} = \mathbf{E}_0 e^{i(kz-\omega t)}$, we can write the sourceless Maxwell equations ($\mu = 1$)

$$\nabla \times \mathbf{E} = -\frac{1}{c}\frac{\partial \mathbf{H}}{\partial t} \qquad (\text{S.}3.44.1)$$

$$\nabla \times \mathbf{H} = \frac{\varepsilon}{c}\frac{\partial \mathbf{E}}{\partial t} \qquad (\text{S.}3.44.2)$$

Substituting the explicit form for $\mathbf{E}$ (and $\mathbf{H}$) produces the following exchange:

$$\frac{\partial}{\partial t} \to -i\omega$$

$$\nabla \times \to i\mathbf{k}\times$$

So (S.3.44.1) and (S.3.44.2) become

$$i\omega \mathbf{H} = ic\mathbf{k} \times \mathbf{E} \qquad (\text{S.}3.44.3)$$

$$i\omega\varepsilon \mathbf{E} = -ic\mathbf{k} \times \mathbf{H} \qquad (\text{S.}3.44.4)$$

Orient the axes so that $\mathbf{E} = E\hat{\mathbf{x}}$ and $\mathbf{H} = H\hat{\mathbf{y}}$ (see Figure S.3.44). Then, the boundary conditions (which require continuity for the tangential components of $\mathbf{E}$ and $\mathbf{H}$) become

$$E_x^{(1)} = E_x^{(2)} \qquad\qquad H_y^{(1)} = H_y^{(2)}$$

where the indices 1 and 2 correspond to dielectric media 1 and 2 (see Figure

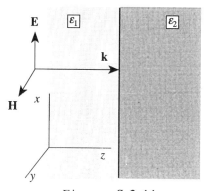

Figure S.3.44

S.3.44). From (S.3.44.3),
$$H_y = \frac{ck}{\omega} E_x \qquad \text{(S.3.44.5)}$$

The field in medium 1 is the sum of the incident wave E_0 and the reflected wave E_1, whereas the field in medium 2 is due only to the transmitted wave E_2. Using the boundary conditions and (S.3.44.5), we obtain

$$E_0 + E_1 = E_2$$
$$k_1(E_0 - E_1) = k_2 E_2 \qquad \text{(S.3.44.6)}$$

Solving (S.3.44.6) for E_1 and E_2,

$$E_1 = \frac{k_1 - k_2}{k_1 + k_2} E_0 = \frac{\sqrt{\varepsilon_1} - \sqrt{\varepsilon_2}}{\sqrt{\varepsilon_1} + \sqrt{\varepsilon_2}} E_0$$
$$E_2 = \frac{2k_1}{k_1 + k_2} E_0 = \frac{2\sqrt{\varepsilon_1}}{\sqrt{\varepsilon_1} + \sqrt{\varepsilon_2}} E_0 \qquad \text{(S.3.44.7)}$$

b) The energy flux in a monochromatic wave is given by the magnitude of the Poynting vector,

$$S = \frac{c}{8\pi} \sqrt{\frac{\varepsilon}{\mu}} \, EE^* \qquad \text{(S.3.44.8)}$$

(see, for instance, Landau and Lifshitz, *Electrodynamics of Continuous Media*, p. 285). So the incident and transmitted fluxes S_0 and S_2, respectively, are, from (S.3.44.7)

$$S_0 = \frac{c}{8\pi} \sqrt{\varepsilon_1} \, |E_0|^2$$

$$S_2 = \frac{c}{8\pi} \sqrt{\varepsilon_2} \, |E_2|^2 = \frac{c}{8\pi} \sqrt{\varepsilon_2} \left(\frac{2\sqrt{\varepsilon_1}}{\sqrt{\varepsilon_1} + \sqrt{\varepsilon_2}} \right)^2 |E_0|^2$$

The fraction of the energy transmitted into the second medium is

$$T = \frac{S_2}{S_0} = \sqrt{\frac{\varepsilon_2}{\varepsilon_1}} \, \frac{4\varepsilon_1}{\left(\sqrt{\varepsilon_1} + \sqrt{\varepsilon_2} \right)^2} = \frac{4\sqrt{\varepsilon_1}\sqrt{\varepsilon_2}}{\left(\sqrt{\varepsilon_1} + \sqrt{\varepsilon_2} \right)^2} = \frac{4n_1 n_2}{(n_1 + n_2)^2} \qquad \text{(S.3.44.9)}$$

where we have substituted $n_{1,2} \equiv \sqrt{\varepsilon_{1,2}}$, the indices of refraction of the two media. Similarly, the fraction of the energy reflected back into the first medium

$$R = \frac{S_1}{S_0} = \left(\frac{n_1 - n_2}{n_1 + n_2} \right)^2 \qquad \text{(S.3.44.10)}$$

where S_1 is the magnitude of the Poynting vector for the reflected wave. We can check that $T + R = 1$ by adding (S.3.44.9) and (S.3.44.10).

3.45   X-Ray Mirror (Princeton)

Calculate the dielectric constant of the metal under the assumption that the electrons in metals are free and disregarding any scattering on the atoms, since they are much heavier than electrons $(M \gg m)$. Under the influence of the X-rays, the electrons only move a small fraction of the distance between atoms, so the field due to the atoms is nearly uniform over one period of the X-ray. Thus, we may write

$$m\dot{\mathbf{v}} = e\mathbf{E}$$

where $\mathbf{E} = \mathbf{E}_0 e^{i(\mathbf{k}\cdot\mathbf{r}-\omega t)}$, or

$$\mathbf{v} = \frac{e}{m}\int \mathbf{E}dt = i\frac{e\mathbf{E}}{m\omega} \tag{S.3.45.1}$$

The current density in the field is

$$\mathbf{J} = en\mathbf{v} = \frac{ie^2 n}{m\omega}\mathbf{E} \tag{S.3.45.2}$$

where n is the density of electrons in the metal. Using the Maxwell equation,

$$\nabla \times \mathbf{H} = \frac{1}{c}\frac{\partial \mathbf{E}}{\partial t} + \frac{4\pi}{c}\mathbf{J} \tag{S.3.45.3}$$

Substituting (S.3.45.2) into (S.3.45.3), we have

$$\nabla \times \mathbf{H} = \left(-\frac{\omega}{c}i + \frac{4\pi}{c}\frac{ne^2}{m\omega}i\right)\mathbf{E} = \frac{1}{c}\frac{\partial \mathbf{D}}{\partial t} = \frac{\varepsilon}{c}\left(-i\omega\right)\mathbf{E} \tag{S.3.45.4}$$

in which we used $\mathbf{D} = \varepsilon\mathbf{E}$. From (S.3.45.4)

$$\varepsilon = 1 - \frac{4\pi ne^2}{m\omega^2} \tag{S.3.45.5}$$

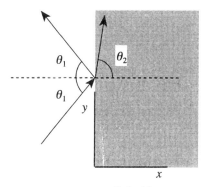

Figure S.3.45

For two media with $\varepsilon_2 < \varepsilon_1$ (see Figure S.3.45),

$$\frac{\sin \theta_2}{\sin \theta_1} = \frac{n_1}{n_2} = \sqrt{\frac{\varepsilon_1}{\varepsilon_2}} > 1$$

For $\theta_2 = \pi/2$, the largest angle of refraction corresponds to the critical angle θ_0 for the incident wave:

$$\sin \theta_0 = \sqrt{\frac{\varepsilon_2}{\varepsilon_1}}$$

So

$$\theta_0 = \sin^{-1} \sqrt{\varepsilon_2/\varepsilon_1} = \sin^{-1} \sqrt{1 - \frac{4\pi n e^2}{m\omega^2}}$$

3.46   Plane Wave in Metal (Colorado, MIT)

a) Starting with Maxwell's equations, we may follow a standard procedure to arrive at the wave equation for the fields and then the dispersion relations:

$$\nabla \cdot \mathbf{D} = 4\pi \rho \qquad \text{(S.3.46.1)}$$

$$\nabla \times \mathbf{E} = -\frac{1}{c} \frac{\partial \mathbf{B}}{\partial t} \qquad \text{(S.3.46.2)}$$

$$\nabla \cdot \mathbf{B} = 0 \qquad \text{(S.3.46.3)}$$

$$\nabla \times \mathbf{H} = \frac{4\pi \mathbf{J}}{c} + \frac{1}{c} \frac{\partial \mathbf{D}}{\partial t} \qquad \text{(S.3.46.4)}$$

First take the curl of (S.3.46.4)

$$\nabla \times \nabla \times \mathbf{H} = \frac{4\pi}{c} \nabla \times \mathbf{J} + \frac{1}{c} \frac{\partial}{\partial t} \nabla \times \mathbf{D} \qquad \text{(S.3.46.5)}$$

Using the identity

$$\nabla \times \nabla \times \mathbf{A} = \nabla \left(\nabla \cdot \mathbf{A} \right) - \nabla^2 \mathbf{A}$$

in (S.3.46.5), we obtain

$$-\nabla^2 \mathbf{H} = \frac{4\pi}{c} \nabla \times \mathbf{J} + \frac{1}{c} \frac{\partial}{\partial t} \nabla \times \mathbf{D} \qquad \text{(S.3.46.6)}$$

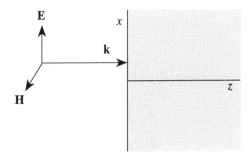

Figure S.3.46

Inside the conductor, we must use the relations $\mathbf{J} = \sigma\mathbf{E}$ and $\mathbf{D} = \varepsilon\mathbf{E}$ in (S.3.46.6). Therefore (S.3.46.6) becomes

$$-\nabla^2\mathbf{H} = \frac{4\pi\sigma}{c}\nabla\times\mathbf{E} + \frac{\varepsilon}{c}\frac{\partial}{\partial t}\nabla\times\mathbf{E} \qquad (\text{S.3.46.7})$$

Using (S.3.46.2) and $\mathbf{B} = \mu\mathbf{H}$ in (S.3.46.7), we obtain the wave equation for $\mathbf{H}$ of the wave propagating in the z direction (see Figure S.3.46):

$$\frac{\partial^2\mathbf{H}}{\partial z^2} - \frac{4\pi\sigma\mu}{c^2}\frac{\partial\mathbf{H}}{\partial t} - \frac{\varepsilon\mu}{c^2}\frac{\partial^2\mathbf{H}}{\partial t^2} = 0 \qquad (\text{S.3.46.8})$$

Disregarding the displacement current (which is equivalent to the condition $\sigma \gg \omega$), we obtain from (S.3.46.8)

$$\frac{\partial^2\mathbf{H}}{\partial z^2} - \frac{4\pi\sigma\mu}{c^2}\frac{\partial\mathbf{H}}{\partial t} = 0 \qquad (\text{S.3.46.9})$$

Substituting the plane wave solution $H \propto \exp i(kz - \omega t)$ into (S.3.46.9) results in

$$-k^2 + i\frac{4\pi\sigma\mu\omega}{c^2} = 0 \qquad (\text{S.3.46.10})$$

or

$$k = \frac{\sqrt{4\pi\sigma\mu\omega}}{c}\sqrt{i} = \frac{\sqrt{2\pi\sigma\mu\omega}}{c}(1 + i) \qquad (\text{S.3.46.11})$$

For our case $\mu = 1$, so we have

$$k = \frac{\sqrt{2\pi\sigma\omega}}{c}(1 + i) \qquad (\text{S.3.46.12})$$

Assuming that the electric field in the incident wave is polarized in the x direction and the magnetic field in the y direction, and the amplitude of the field outside the conductor is E_0, we can obtain the fields inside $\mathbf{H}_t$, $\mathbf{E}_t$

$$\mathbf{H}_t = H_t\hat{\mathbf{y}}e^{i(kz-\omega t)} = H_t\hat{\mathbf{y}}e^{-z/\delta+i(z/\delta-\omega t)}$$

where $\delta \equiv c/\sqrt{2\pi\sigma\omega}$ is the characteristic penetration depth of the field (skin depth). From the boundary conditions $H_t^{in} = H_t^{out} = E_0$, we have

$$\mathbf{H}_t = E_0\hat{\mathbf{y}}e^{-z/\delta+i(z/\delta-\omega t)} \qquad (S.3.46.13)$$

The electric field inside the conductor $\mathbf{E}_t$:

$$\mathbf{E}_t = E_t\hat{\mathbf{x}} = \frac{c}{4\pi\sigma}\nabla \times \mathbf{H}_t = -\frac{c}{4\pi\sigma}\frac{\partial H_t}{\partial z}\hat{\mathbf{x}}$$

$$= -\frac{c}{4\pi\sigma}ikE_0\hat{\mathbf{x}}e^{-z/\delta+i(z/\delta-\omega t)} = \frac{c}{4\pi\sigma}\frac{\sqrt{2\pi\sigma\omega}}{c}(1-i)E_0\hat{\mathbf{x}}e^{-z/\delta+i(z/\delta-\omega t)}$$

$$= \sqrt{\frac{\omega}{8\pi\sigma}}(1-i)E_0\hat{\mathbf{x}}e^{-z/\delta+i(z/\delta-\omega t)} \qquad (S.3.46.14)$$

Therefore

$$\mathbf{E}_t = \sqrt{\frac{\omega}{4\pi\sigma}}E_0\hat{\mathbf{x}}e^{-z/\delta+i(z/\delta-\omega t-\pi/4)} \qquad (S.3.46.15)$$

where the phase shift $-\pi/4$ comes from the factor $(1-i)/\sqrt{2}$. Equation (S.3.46.14) is a special case of a more general formula

$$\mathbf{E}_t = \frac{c}{4\pi}Z_s\left(\mathbf{H}_t \times \mathbf{n}\right)$$

where $\mathbf{n}$ is a unit vector in the direction of the wave propagation, and

$$Z_s \equiv \frac{1}{c}\sqrt{\frac{2\pi\omega}{\sigma}}(1-i)$$

is the surface resistance.

b) The ratio of the amplitude of the magnetic field to that of the electric field inside the metal from (S.3.46.13) and (S.3.46.15) is in this approximation

$$\frac{|\mathbf{H}_t|}{|\mathbf{E}_t|} = \frac{1}{\sqrt{\omega/4\pi\sigma}} = \sqrt{\frac{4\pi\sigma}{\omega}} = 2\sqrt{\frac{\pi\sigma}{\omega}}$$

Therefore the energy of the field inside a good conductor is mostly the magnetic energy.

c) The power per unit area P transmitted into the metal is given by the flux of the Poynting vector:

$$P = -\frac{c}{8\pi}\text{Re}\left\{\mathbf{E} \times \mathbf{H}^* \cdot \mathbf{n}\right\}\Big|_{z=0} = \frac{c}{16\pi}\sqrt{\frac{\omega}{2\pi\sigma}}E_0^2$$

3.47 Wave Attenuation (Stony Brook)

a,b) We obtain the equation for the electric (magnetic) field in the same way as in Problem 3.46 (see (S.3.46.8))

$$\nabla^2\mathbf{E} - \frac{\varepsilon\mu}{c^2}\frac{\partial^2\mathbf{E}}{\partial t^2} - \frac{4\pi\sigma\mu}{c^2}\frac{\partial\mathbf{E}}{\partial t} = 0 \qquad (S.3.47.1)$$

c) Now, taking E_y in the form $E_y = \psi_0 e^{i(kx-\omega t)}$ and substituting into (S.3.47.1) yields

$$-k^2 + \frac{\varepsilon\mu\omega^2}{c^2} + i\frac{4\pi\sigma\mu\omega}{c^2} = 0 \qquad (S.3.47.2)$$

We have

$$k = \sqrt{\varepsilon\mu}\,\frac{\omega}{c}\left(1 + i\frac{4\pi\sigma}{\varepsilon\omega}\right)^{1/2} \equiv k' + k''i \qquad (S.3.47.3)$$

To solve for the square root of the complex expression in (S.3.47.3), write

$$\left(1 + i\frac{4\pi\sigma}{\varepsilon\omega}\right)^{1/2} \equiv a + bi \qquad (S.3.47.4)$$

where a and b are real, and

$$k' = \sqrt{\varepsilon\mu}\,\frac{\omega}{c}a$$

$$k'' = \sqrt{\varepsilon\mu}\,\frac{\omega}{c}b$$

By squaring (S.3.47.4), we find

$$a^2 - b^2 = 1 \qquad (S.3.47.5)$$

$$2ab = \frac{4\pi\sigma}{\varepsilon\omega} \qquad (S.3.47.6)$$

Taking a from (S.3.47.6) and substituting it into (S.3.47.5), we have for b

$$b = \left(-\frac{1}{2} + \sqrt{\frac{1}{4} + \left(\frac{2\pi\sigma}{\varepsilon\omega} \right)^2} \right)^{1/2}$$

where we have chosen the branch of the root with the plus sign to satisfy $\sigma = 0$ (vacuum) $b = k'' = 0$ (no dissipation). So

$$k'' = \sqrt{\frac{\varepsilon\mu}{2}} \frac{\omega}{c} \left(\sqrt{\left(\frac{4\pi\sigma}{\varepsilon\omega} \right)^2 + 1} - 1 \right)^{1/2} \qquad \text{(S.3.47.7)}$$

Therefore, the attenuation length ($1/e$ point) for the amplitude

$$\delta = (k'')^{-1} = \sqrt{\frac{2}{\varepsilon\mu}} \frac{c}{\omega} \left(\sqrt{\left(\frac{4\pi\sigma}{\varepsilon\omega} \right)^2 + 1} - 1 \right)^{-1/2} \qquad \text{(S.3.47.8)}$$

whereas for the intensity, the attenuation length is $\delta/2$.

d) For the frequency given in the problem,

$$\frac{4\pi\sigma}{\varepsilon\omega} \gg 1$$

and so we can disregard the 1's in (S.3.47.8) and rewrite it as

$$\delta \approx \sqrt{\frac{2}{\varepsilon\mu}} \frac{c}{\omega} \frac{\sqrt{\varepsilon\omega}}{2\sqrt{\pi\sigma}} = \frac{c}{\sqrt{2\pi\omega\sigma}} = \frac{3 \cdot 10^{10}}{\sqrt{2 \cdot 3.14 \cdot 5 \cdot 10^5 \cdot 4.5 \cdot 10^{10}}} \text{ cm} \approx 80 \text{ cm}$$

At a depth of 10 m below the surface, the intensity attenuation at this frequency will be

$$\frac{P}{P_0} \approx \exp\left[-(2 \cdot 10/0.8) \right] \approx 10^{-11}$$

which implies that transmission of signals to submerged submarines will require much lower frequencies, $f \leq 10\text{--}10^2 \text{Hz}$ (see also Jackson, *Classical Electrodynamics* §7.5).

3.48 Electrons and Circularly Polarized Waves (Boston)

a) Find the equations of motion of the electron in the x and y directions at $z = 0$ for $\varepsilon = -1$:

$$m\ddot{x} = -m\omega_0^2 x - eE_0 \cos(\omega t) - \frac{e}{c} B\dot{y} \qquad (S.3.48.1)$$

$$m\ddot{y} = -m\omega_0^2 y + eE_0 \sin(\omega t) + \frac{e}{c} B\dot{x} \qquad (S.3.48.2)$$

Multiply (S.3.48.2) by $-i$ and add to (S.3.48.1). Substituting the variable η from the hint in the problem, we obtain

$$m(\ddot{\eta} + \omega_0^2 \eta) + \frac{ieB\dot{\eta}}{c} = -eE_0 e^{i\omega t} \qquad (S.3.48.3)$$

(S.3.48.3) may be solved by finding a solution at the driving frequency ω:

$$\eta = Ae^{i\omega t} \qquad A = -\frac{eE_0/m}{\omega_0^2 - \omega^2 - eB\omega/mc}$$

The analogous differential equation for ξ is formed by multiplying (S.3.48.2) by i and adding to (S.3.48.1). We find that $\xi = A\exp(-i\omega t)$. This gives for the original variables

$$x = A\cos(\omega t)$$

$$y = -A\sin(\omega t)$$

For $\varepsilon = +1$, we have

$$\xi = A'e^{i\omega t} \qquad \eta = A'e^{-i\omega t}$$

where

$$A' = -\frac{eE_0/m}{\omega_0^2 - \omega^2 + eB\omega/mc}$$

Here,

$$x = A'\cos(\omega t)$$

$$y = -A'\sin(\omega t)$$

b) A linearly polarized wave may be viewed as the sum of two circularly polarized waves of opposite helicity (see Figure S.3.48). After propagating through the medium, the waves will be delayed by different phases, given

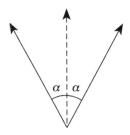

Figure S.3.48

by α_+ and α_-

$$\alpha_+ = \alpha + k_+ l = \alpha + \frac{n_+ \omega l}{c}$$

$$\alpha_- = \alpha + \frac{n_- \omega l}{c}$$

The plane of polarization is now along the bisector of the circularly polarized vectors and forms with the initial plane of polarization the angle

$$\chi = -\alpha_+ + \frac{\alpha_+ + \alpha_-}{2} = \frac{\alpha_- - \alpha_+}{2} = -\frac{\omega l}{2c} (n_+ - n_-) \quad \text{(S.3.48.4)}$$

$$= -\left(\frac{\omega l}{2c}\right) \frac{n_+^2 - n_-^2}{n_+ + n_-} \approx -\left(\frac{\omega l}{2c}\right) \frac{n_+^2 - n_-^2}{2n} \quad \text{(S.3.48.5)}$$

where $2n \approx n_+ + n_-$. Substituting (P.3.48.1) into (S.3.48.5), we obtain

$$\chi \approx \frac{\omega l}{2c} \frac{1}{2n} \frac{4\pi e^2 N}{m} \frac{2eB\omega/mc}{\left(\omega_0^2 - \omega^2\right)^2} = \frac{4\pi e^2 N}{m} \frac{1}{\left(\omega_0^2 - \omega^2\right)^2} \frac{eB\omega^2}{mc^2} \frac{l}{2n} \quad \text{(S.3.48.6)}$$

3.49   Classical Atomic Spectral Line (Princeton, Wisconsin-Madison)

a) The equation

$$m\ddot{\mathbf{x}} + m\omega_0^2 \mathbf{x} + \gamma\dot{\mathbf{x}} = 0$$

has a solution

$$\mathbf{x} \approx \text{Re}\left\{\mathbf{x}_0 e^{-(\gamma/2m)t} e^{i\omega_0 t}\right\} \quad \text{(S.3.49.1)}$$

with initial conditions $\mathbf{x} = \mathbf{x}_0$, and $\dot{\mathbf{x}} = 0$, where we used

$$\sqrt{\left(\frac{\gamma}{2m}\right)^2 - \omega_0^2} \approx i\omega_0$$

for $\omega_0 \gg \gamma/m$. In the same approximation, the acceleration

$$|\mathbf{w}| = |\ddot{\mathbf{x}}| = \left| \mathrm{Re} \left\{ -\mathbf{x}_0 \omega_0^2 e^{-(\gamma/2m)t} e^{i\omega_0 t} \right\} \right| \propto |\mathbf{x}| \qquad (\mathrm{S.3.49.2})$$

The total energy $\mathcal{E}$ radiated by the atom

$$\mathcal{E} = \int_0^\infty I(\omega) d\omega \qquad (\mathrm{S.3.49.3})$$

where $I(\omega)$ is the spectral density of the radiation. On the other hand

$$\mathcal{E} = \int_0^\infty P(t)\, dt \propto \int_0^\infty w^2\, dt = \int_{-\infty}^\infty w^2\, dt \qquad (\mathrm{S.3.49.4})$$

where we used the formula

$$P = \frac{2}{3} \frac{e^2 w^2}{c^3} \qquad (\mathrm{S.3.49.5})$$

and the fact that $w = 0$ at $t < 0$. We may write the Fourier transform of the acceleration w:

$$W(\omega) \propto \int_{-\infty}^\infty w(t) e^{-i\omega t}\, dt \propto \int_{-\infty}^\infty |\mathbf{x}|\, e^{-i\omega t}\, dt$$

$$\propto \int_0^\infty e^{-(\gamma/2m)t + i\omega_0 t - i\omega t}\, dt = \frac{1}{(\gamma/2m) + i(\omega - \omega_0)}$$

Now, Parseval's relation (see, for instance, Arfken, *Mathematical Methods for Physicists*) gives

$$\mathcal{E} \propto \int_{-\infty}^\infty w^2(t)\, dt = \int_{-\infty}^\infty |W(\omega)|^2\, d\omega \propto \int_{-\infty}^\infty \frac{d\omega}{(\gamma/2m)^2 + (\omega - \omega_0)^2} \qquad (\mathrm{S.3.49.6})$$

Comparing (S.3.49.3) and (S.3.49.6), we obtain for the spectral density

$$I(\omega) \propto \frac{1}{(\gamma/2m)^2 + (\omega - \omega_0)^2}$$

b) The energy U of the oscillator may be written

$$U = U_0 e^{-\Gamma t} = \frac{1}{2} m \omega_0^2 x_0^2 e^{-\Gamma t} \qquad (S.3.49.7)$$

The power loss is therefore

$$\frac{dU}{dt} = -\frac{\Gamma}{2} m \omega_0^2 x_0^2 e^{-\Gamma t} \qquad (S.3.49.8)$$

This may be equated to the power loss given by the average over one cycle of (S.3.49.5)

$$\frac{2e^2 x_0^2 \omega_0^4}{3c^3} \frac{1}{2} = \frac{\Gamma}{2} m \omega_0^2 x_0^2 \qquad (S.3.49.9)$$

$$\Gamma = \frac{2e^2 \omega_0^2}{3mc^3}$$

c) We may rewrite (S.3.49.9) as

$$\Gamma = \frac{8\pi^2 e^2}{3m\lambda^2 c}$$

The linewidth in angstroms may be found from

$$\Gamma = \Delta\omega = \frac{2\pi c \Delta\lambda}{\lambda^2}$$

$$\Delta\lambda = \frac{\lambda^2 \Gamma}{2\pi c} = \frac{\lambda^2}{2\pi c} \frac{8\pi^2 e^2}{3m\lambda^2 c} = \frac{4\pi e^2}{3mc^2}$$

$$\approx \frac{4 \cdot 3.14 \cdot \left(4.8 \cdot 10^{-10}\right)^2}{3 \cdot (0.91 \cdot 10^{-27})(3 \cdot 10^{10})^2} \approx 10^{-12} \text{ cm} = 10^{-4} \text{ Å}$$

Now find the time T for the atom to lose half its energy:

$$T = \frac{\ln 2}{\Gamma}$$

The number of oscillations is then

$$fT = \frac{c \ln 2}{\lambda \Gamma} = \frac{c \ln 2}{\lambda} \frac{3m\lambda^2 c\hbar}{8\pi^2 e^2 \hbar}$$

$$= \frac{c}{\hbar} \frac{\ln 2}{\alpha} \frac{3m\lambda}{8\pi^2} \approx \frac{3 \cdot 10^{10}}{1 \cdot 10^{-27}} \frac{\ln 2}{1/137} \frac{3 \cdot 1 \cdot 10^{-27} \cdot 5 \cdot 10^{-5}}{8 \cdot \pi^2}$$

$$\approx 5 \cdot 10^6 \text{oscillations}$$

3.50   Lifetime of Classical Atom (MIT, Princeton, Stony Brook)

If the energy loss per revolution is small compared to the total energy of the electron in the atom, we can write (see Problem 3.49)

$$\frac{d\mathcal{E}}{dt} = -P_{\text{rad}} = -\frac{2}{3}\frac{e^2 w^2}{c^3} \tag{S.3.50.1}$$

where w is the acceleration of the electron and P_{rad} is the total radiated power. Using our assumption, we can approximate the orbit of the electron (which is a spiral) by a circle for each revolution of radius $r = r(t)$. The acceleration is due to the Coulomb force

$$w = \frac{e^2}{mr^2} \tag{S.3.50.2}$$

On the other hand, using $|\mathcal{E}| = |U|/2$ (U is the potential energy of a particle moving in a circle in a $1/r^2$ field; see Problem P.1.44) we have

$$\mathcal{E} = -\frac{1}{2}\frac{e^2}{r} \tag{S.3.50.3}$$

Substituting (S.3.50.2) and (S.3.50.3) into (S.3.50.1) gives

$$\frac{1}{2}\frac{e^2}{r^2}\frac{dr}{dt} = -\frac{2}{3}\frac{e^2}{c^3}\frac{e^4}{m^2 r^4} \tag{S.3.50.4}$$

or

$$\frac{dr}{dt} = -\frac{4}{3}\frac{e^4}{m^2 c^3 r^2} \tag{S.3.50.5}$$

Integrating (S.3.50.5) yields

$$t \approx -\frac{3m^2 c^3}{4e^4}\int_{a_0}^{0} r^2\, dr = \frac{3}{4}\frac{m^2 c^3}{e^4}\frac{a_0^3}{3} = \frac{m^2 c^3 a_0^3}{4e^4}$$

So

$$t = \frac{(9\cdot 10^{-28})^2(3\cdot 10^{10})^3(0.5\cdot 10^{-8})^3}{4(4.8\cdot 10^{-10})^4} \approx 10^{-11}\text{ s}$$

3.51   Lorentz Transformation of Fields (Stony Brook)

a) Using a 4-vector of the form

$$A^\mu = (A^0, \mathbf{A}) \qquad \text{(contravariant form)}$$

and recalling that for $\mathbf{v}$ pointing along the x-axis, we have for the Lorentz transformation of a 4-vector

$$A^{0'} = \gamma \left(A^0 - \frac{v}{c} A^1 \right) \tag{S.3.51.1}$$

$$A^{1'} = \gamma \left(A^1 - \frac{v}{c} A^0 \right) \tag{S.3.51.2}$$

$$A^{2'} = A^2 \tag{S.3.51.3}$$

$$A^{3'} = A^3 \tag{S.3.51.4}$$

For space–time coordinates,

$$A^\mu = (ct, \mathbf{r})$$

and we have from (S.3.51.1)–(S.3.51.4)

$$t' = \gamma(t - \frac{v}{c^2} x) \tag{S.3.51.5}$$

$$x' = \gamma(x - vt) \tag{S.3.51.6}$$

$$y' = y \tag{S.3.51.7}$$

$$z' = z \tag{S.3.51.8}$$

b) Writing the explicitly antisymmetric field-strength tensor,

$$\mathcal{F}^{\mu\nu} = \begin{pmatrix} 0 & -E_x & -E_y & -E_z \\ E_x & 0 & -B_z & B_y \\ E_y & B_z & 0 & -B_x \\ E_z & -B_y & B_x & 0 \end{pmatrix} \tag{S.3.51.9}$$

where $\mathcal{F}^{01}$ and $\mathcal{F}^{23}$ do not change under the Lorentz transformation and $\mathcal{F}^{02}$, $\mathcal{F}^{03}$ and $\mathcal{F}^{12}$, $\mathcal{F}^{13}$ transform as x^0 and x^1, respectively (see, for instance, Landau and Lifshitz, *Classical Theory of Fields*, Chapter 1.6).

$$F^{02'} = \gamma \left(F^{02} - \frac{v}{c} F^{12} \right)$$

$$F^{03'} = \gamma \left(F^{03} - \frac{v}{c} F^{13} \right)$$

$$F^{12'} = \gamma \left(F^{12} - \frac{v}{c} F^{02} \right) \qquad \text{(S.3.51.10)}$$

$$F^{13'} = \gamma \left(F^{13} - \frac{v}{c} F^{03} \right)$$

Substituting (S.3.51.9) into (S.3.51.10), we obtain

$$E'_x = E_x \quad E'_y = \gamma \left(E_y - \frac{v}{c} B_z \right) \quad E'_z = \gamma \left(E_z + \frac{v}{c} B_y \right)$$

$$\text{(S.3.51.11)}$$

$$B'_x = B_x \quad B'_y = \gamma \left(B_y + \frac{v}{c} E_z \right) \quad B'_z = \gamma \left(B_z - \frac{v}{c} E_y \right)$$

We can rewrite (S.3.51.11) is terms of the parallel and perpendicular components of the fields:

$$\mathbf{E}'_\parallel = E_x \hat{\mathbf{x}} = \mathbf{E}_\parallel \quad \mathbf{E}'_\perp = E'_y \hat{\mathbf{y}} + E'_z \hat{\mathbf{z}} = \gamma \left(\mathbf{E}_\perp + \frac{\mathbf{v}}{c} \times \mathbf{B} \right)$$

$$\text{(S.3.51.12)}$$

$$\mathbf{B}'_\parallel = B_x \hat{\mathbf{x}} = \mathbf{B}_\parallel \quad \mathbf{B}'_\perp = B'_y \hat{\mathbf{y}} + B'_z \hat{\mathbf{z}} = \gamma \left(\mathbf{B}_\perp - \frac{\mathbf{v}}{c} \times \mathbf{E} \right)$$

c) In the case of a point charge q, we have to transform from K' to K, which is equivalent to changing the sign of the velocity in (S.3.51.12). For a small velocity $\mathbf{v}$, we may write

$$\mathbf{B}_\perp \approx \mathbf{B}'_\perp + \frac{\mathbf{v}}{c} \times \mathbf{E}'$$

where we have changed the signs in (S.3.51.12) and taken $\gamma \approx 1$. For a point charge in K', $\mathbf{B}' = 0$ and

$$\mathbf{B}_\perp = \frac{\mathbf{v}}{c} \times \mathbf{E}' = \frac{q}{c} \frac{\mathbf{v} \times \mathbf{r}}{r^3}$$

which is the magnetic field for a charge moving with velocity $\mathbf{v}$.

3.52   Field of a Moving Charge (Stony Brook)

a) Differentiating, we obtain

$$\frac{1}{c}\frac{\partial\varphi}{\partial t} = \frac{vq_1(z-vt)}{cs^3} \qquad\qquad \text{(S.3.52.1)}$$

and, since $\mathbf{v}$ lies only in the z direction (see Figure S.3.52)

$$\nabla\cdot\mathbf{A} = \frac{\partial A_z}{\partial z} = -\frac{vq_1(z-vt)}{cs^3} \qquad\qquad \text{(S.3.52.2)}$$

b) To calculate $\mathbf{B}$, we recall from the definition of $\mathbf{A}$ that

$$\mathbf{B} \equiv \nabla\times\mathbf{A} = \nabla\times\frac{\mathbf{v}}{c}\varphi = -\frac{\mathbf{v}}{c}\times\nabla\varphi = \frac{\mathbf{v}}{c}\times(-\nabla\varphi)$$

$$= \frac{\mathbf{v}}{c}\times\left(-\nabla\varphi - \frac{1}{c}\frac{\partial\mathbf{A}}{\partial t}\right) = \frac{\mathbf{v}}{c}\times\mathbf{E} \qquad\qquad \text{(S.3.52.3)}$$

where we have used the fact that

$$\mathbf{v}\times\frac{1}{c}\frac{\partial\mathbf{A}}{\partial t} = 0$$

Now,

$$\mathbf{E} \equiv -\nabla\varphi - \frac{1}{c}\frac{\partial\mathbf{A}}{\partial t}$$

$$= q_1\frac{\left(1-\beta^2\right)(x\hat{\mathbf{x}}+y\hat{\mathbf{y}}) + (z-vt)\hat{\mathbf{z}}}{s^3} - \frac{vq_1(z-vt)v}{c^2s^3}$$

$$= \frac{q_1\left[\left(1-\beta^2\right)(x\hat{\mathbf{x}}+y\hat{\mathbf{y}}) + (z-vt)\hat{\mathbf{z}} - v^2/c^2(z-vt)\hat{\mathbf{z}}\right]}{s^3}$$

$$= \frac{q_1\left(1-\beta^2\right)\mathbf{R}}{s^3} = \frac{q_1\mathbf{R}}{\gamma^2 s^3} \qquad\qquad \text{(S.3.52.4)}$$

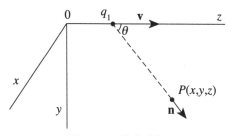

Figure S.3.52

where $\mathbf{R} = x\hat{\mathbf{x}} + y\hat{\mathbf{y}} + (z - vt)\hat{\mathbf{z}}$ is the vector from the charge to point P and therefore is parallel to $\mathbf{n}$. The same results can be obtained by calculating $\mathbf{E}$ in the moving frame (with the charge q_1) and then taking the Lorentz transformation.

c) The force acting on charge q_2 can be calculated as the force acting on q_2 in the field of q_1 from (S.3.52.4):

$$\mathbf{F} = q_2\mathbf{E} + q_2\frac{\mathbf{v}}{c} \times \mathbf{B} = q_2\mathbf{E} + \frac{q_2}{c^2}\mathbf{v} \times (\mathbf{v} \times \mathbf{E}) \qquad (\text{S.3.52.5})$$

where we used (S.3.52.3). Substituting $\mathbf{E}$ from (S.3.52.4), we find

$$\mathbf{F} = \frac{q_1q_2\mathbf{R}}{\gamma^2 s^3} + \frac{q_1q_2}{c^2}\mathbf{v} \times \left(\mathbf{v} \times \frac{\mathbf{R}}{\gamma^2 s^3}\right)$$

$$= \frac{q_1q_2}{\gamma^2 s^3}\left(1 - \frac{v^2}{c^2}\right)\mathbf{R} + \frac{q_1q_2\,(\mathbf{R} \cdot \mathbf{v})\,\mathbf{v}}{c^2\gamma^2 s^3}$$

We can express s through the angle θ between $\mathbf{R}$ and $\hat{\mathbf{z}}$ (see Figure S.3.52):

$$s^2 = R^2\left(1 - \beta^2\sin^2\theta\right)$$

where we used $x^2 + y^2 = R^2\sin^2\theta$, and so $s = |\mathbf{R}|\left(1 - \beta^2\sin^2\theta\right)^{1/2}$. Now, $\mathbf{F}$ may be written as a sum of the projections perpendicular and parallel to the z direction:

$$F_\perp = \frac{q_1q_2}{\gamma^2 R^3}\frac{(1 - v^2/c^2)R\sin\theta}{\left(1 - \beta^2\sin^2\theta\right)^{3/2}} = \frac{q_1q_2}{R^2}\frac{\sin\theta}{\gamma^4\left(1 - \beta^2\sin^2\theta\right)^{3/2}}$$

$$F_\parallel = \frac{q_1q_2(1 - v^2/c^2)R\cos\theta}{\gamma^2 R^3(1 - \beta^2\sin^2\theta)^{3/2}} + \frac{q_1q_2Rv^2\cos\theta}{c^2\gamma^2 R^3(1 - \beta^2\sin^2\theta)^{3/2}}$$

$$= \frac{q_1q_2}{R^2}\frac{\cos\theta}{\gamma^2(1 - \beta^2\sin^2\theta)^{3/2}}$$

3.53 Retarded Potential of Moving Line Charge (MIT)

a) We may calculate the field of a line charge using Gauss's law

$$\int_S \mathbf{E} \cdot d\mathbf{A} = 4\pi q$$

$$E \cdot 2\pi r l = 4\pi\lambda l$$

where r is the distance from the line charge and l is some length of wire. So

$$E_r = \frac{2\lambda}{r} \tag{S.3.53.1}$$

b) The current density

$$\mathbf{J} = \lambda v \delta(x)\delta(y)\theta(t)\hat{\mathbf{z}} \tag{S.3.53.2}$$

where δ is the Dirac delta function and $\theta(t)$ is defined by

$$\theta(t) = \begin{cases} 0 & t < 0 \\ 1 & t \geq 0 \end{cases}$$

We may then write

$$A_z(x_0, t)$$

$$= \frac{1}{c} \int \int \int dx'dy'dz'\lambda v \delta(x')\delta(y') \frac{\theta\left(t - (1/c)\sqrt{(x_0 - x')^2 + y'^2 + z'^2}\right)}{\sqrt{(x_0 - x')^2 + y'^2 + z'^2}}$$

$$= \frac{\lambda v}{c} \int\limits_{z'=-\infty}^{\infty} \frac{\theta\left(t - (1/c)\sqrt{x_0^2 + z'^2}\right) dz'}{\sqrt{x_0^2 + z'^2}} \tag{S.3.53.3}$$

Now, $\theta\left(t - (1/c)\sqrt{x_0^2 + z'^2}\right)$ is zero unless $t > (1/c)\sqrt{x_0^2 + z'^2}$, so

$$|z'| < \sqrt{c^2t^2 - x_0^2}$$

and the integral in (S.3.53.3) becomes

$$A_z(x_0, t) = \frac{\lambda v}{c} \int\limits_{-\sqrt{c^2t^2 - x_0^2}}^{\sqrt{c^2t^2 - x_0^2}} \frac{dz'}{\sqrt{x_0^2 + z'^2}}$$

$$= \frac{\lambda v}{c} \ln \left| \frac{ct + \sqrt{c^2t^2 - x_0^2}}{ct - \sqrt{c^2t^2 - x_0^2}} \right| = \frac{\lambda v}{c} \ln \left| \frac{1 + \sqrt{1 - (x_0/ct)^2}}{1 - \sqrt{1 - (x_0/ct)^2}} \right| \tag{S.3.53.4}$$

for $t > x_0/c$. For $t < x_0/c$, $A_z(x_0, t) = 0$.

c) From (S.3.53.4), we have for $\rho < ct$

$$A_z(\rho, t) = \frac{\lambda v}{c} \ln \left| \frac{1 + \sqrt{1 - (\rho/ct)^2}}{1 - \sqrt{1 - (\rho/ct)^2}} \right| \qquad (S.3.53.5)$$

By definition, $\mathbf{B} \equiv \nabla \times \mathbf{A}$, which in cylindrical coordinates gives (see Appendix)

$$B_\varphi = -\frac{\partial}{\partial \rho} A_z(\rho, t) = -\frac{\lambda v}{c} \left[-\frac{\rho/(ct)^2}{\sqrt{1 - (\rho/ct)^2} \left(1 + \sqrt{1 - (\rho/ct)^2} \right)} \right.$$

$$\left. - \frac{\rho/(ct)^2}{\sqrt{1 - (\rho/ct)^2} \left(1 - \sqrt{1 - (\rho/ct)^2} \right)} \right] = \frac{\lambda v}{c} \frac{2}{\rho \sqrt{1 - (\rho/ct)^2}}$$

$$\approx \frac{\lambda v}{c} \frac{2}{\rho} = \frac{2I}{c\rho}$$

for $t \to \infty$, which is the value of the magnetic field that would result from a calculation using Ampère's law.

3.54 Orbiting Charges and Multipole Radiation (Princeton, Michigan State, Maryland)

a,b) At $r \gg d$, the emitted radiation is confined to a dipole ($r \gg \lambda \gg d$), where λ is the wavelength. The vector potential of the system with dipole moment $\mathbf{p}$ at a distance $r \gg \lambda$ is given by

$$\mathbf{A} = \frac{1}{cr_0} \dot{\mathbf{p}}$$

The magnetic field of the system (see, for instance, Landau and Lifshitz, *Classical Theory of Fields*)

$$\mathbf{H} = \frac{1}{c} \left[\dot{\mathbf{A}} \times \mathbf{n} \right] = \frac{1}{c^2 r_0} \ddot{\mathbf{p}} \times \mathbf{n} \qquad (S.3.54.1)$$

where $\mathbf{p} = q\mathbf{d}$ is the dipole moment of the system, $\mathbf{n}$ is the unit vector in the direction of observation, and r_0 is the distance from the origin (see

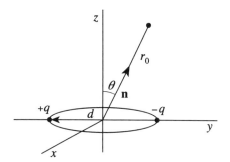

Figure S.3.54a

Figure S.3.54a). The energy flux is given by the Poynting vector $\mathbf{S}$:

$$\mathbf{S} = c\frac{H^2}{4\pi}\mathbf{n} \qquad (S.3.54.2)$$

The radiated power in a solid angle $d\Theta$ is given by

$$dP = Sr_0^2 d\Theta = c\frac{H^2 r_0^2}{4\pi}\,d\Theta \qquad (S.3.54.3)$$

Substituting (S.3.54.1) into (S.3.54.3), we obtain

$$dP = \frac{1}{4\pi c^3}\,|\ddot{\mathbf{p}} \times \mathbf{n}|^2\,d\Theta \qquad (S.3.54.4)$$

Noting that
$$\mathbf{p} = q\mathbf{d} = (p\cos\omega t, p\sin\omega t) = (p_x, p_y)$$

we have

$$\ddot{\mathbf{p}} \times \mathbf{n} = -\begin{vmatrix} \hat{\mathbf{i}} & \hat{\mathbf{j}} & \hat{\mathbf{k}} \\ p_x & p_y & 0 \\ n_x & n_y & n_z \end{vmatrix}\omega^2 = \left[-p_y n_z\hat{\mathbf{i}} + p_x n_z\hat{\mathbf{j}} - (p_x n_y - p_y n_x)\hat{\mathbf{k}}\right]\omega^2$$

$$\left\langle|\ddot{\mathbf{p}} \times \mathbf{n}|^2\right\rangle = \left\langle p^2\cos^2\theta + p^2\cos^2\omega t\,\sin^2\theta\,\sin^2\varphi + p^2\sin^2\omega t\,\sin^2\theta\,\cos^2\varphi\right.$$

$$\left. - 2p^2\sin\omega t\,\cos\omega t\,\sin^2\theta\,\sin\varphi\,\cos\varphi\right\rangle\omega^4$$

$$= \frac{1}{2}p^2(1 + \cos^2\theta)\omega^4$$

where we took the average over the period of revolution and used $\left\langle\sin^2\omega t\right\rangle =$

$\langle \cos^2 \omega t \rangle = 1/2$ and $\langle 2 \sin \omega t \cos \omega t \rangle = \langle \sin 2\omega t \rangle = 0$. So from (S.3.54.4)

$$\langle dP \rangle = \frac{1}{4\pi c^3} \frac{1}{2} p^2 \omega^4 \left(1 + \cos^2 \theta \right) d\Theta = \frac{p^2 \omega^4}{8\pi c^3} \left(1 + \cos^2 \theta \right) d\Theta$$

c) The total power radiated is

$$\langle P \rangle = \int \frac{dP}{d\Theta} \, d\Theta = \frac{p^2 \omega^4}{8\pi c^3} \, 2\pi \int_0^\pi \left(1 + \cos^2 \theta \right) \sin \theta \, d\theta$$

$$= \frac{p^2 \omega^4}{4c^3} \left[-\cos \theta - \frac{1}{3} \cos^3 \theta \right]\Big|_0^\pi = \frac{p^2 \omega^4}{4c^3} \left[2 + \frac{2}{3} \right]$$

$$= \frac{2p^2 \omega^4}{3c^3} = \frac{2q^2 d^2 \omega^4}{3c^3}$$

d) When the plane $z = -b$ is filled with a perfect conductor, we have an image charge for each of the charges $+q$ and $-q$, and the total dipole moment of the system becomes zero (see Figure S.3.54b). The next nonzero multipole in this system must be quadrupole with quadrupole moment $D_{\alpha\beta}$:

$$D_{\alpha\beta} = \sum q_\pm \left(3x_\alpha x_\beta - r^2 \delta_{\alpha\beta} \right)$$

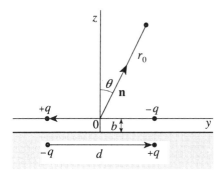

Figure S.3.54b

3.55   Electron and Radiation Reaction (Boston)

a) By assuming that $v \ll c$, we may write

$$m\dot{\mathbf{v}} \approx q \left(\mathbf{E} + \frac{1}{c} \mathbf{v} \times \mathbf{B} \right) \tag{S.3.55.1}$$

Differentiating (S.3.55.1) with respect to time, we obtain

$$\ddot{\mathbf{v}} \approx \frac{q}{m}\dot{\mathbf{E}} + \frac{q}{mc}\dot{\mathbf{v}} \times \mathbf{B} + \frac{q}{mc}\mathbf{v} \times \dot{\mathbf{B}} \qquad (S.3.55.2)$$

Substituting for $\dot{\mathbf{v}}$ in (S.3.55.2) results in

$$\ddot{\mathbf{v}} \approx \frac{q}{m}\left(\dot{\mathbf{E}} + \frac{1}{c}\mathbf{v} \times \dot{\mathbf{B}}\right) + \frac{q^2}{m^2c}\left(\mathbf{E} \times \mathbf{B} + \frac{1}{c}(\mathbf{v} \times \mathbf{B}) \times \mathbf{B}\right)$$

$$= \frac{q}{m}\dot{\mathbf{E}} + \frac{q^2}{m^2c}\mathbf{E} \times \mathbf{B}$$

where we have disregarded terms first order in v/c. So

$$\mathbf{F}_r = \frac{2q^2}{3c^3}\ddot{\mathbf{v}} = \frac{2q^2}{3c^3}\left[\frac{q}{m}\dot{\mathbf{E}} + \frac{q^2}{m^2c}\mathbf{E} \times \mathbf{B}\right] \qquad (S.3.55.3)$$

b) Let the $\mathbf{E}$ field of the plane wave be polarized in the x direction, so that

$$E_x = E_0 \sin(kz - \omega t) \qquad E_y = E_z = 0$$

$$B_y = B_0 \sin(kz - \omega t) \qquad B_x = B_z = 0$$

The time averages of (S.3.55.3) are

$$\left\langle \dot{\mathbf{E}} \right\rangle = 0 \qquad \langle \mathbf{E} \times \mathbf{B} \rangle = \frac{1}{2}E_0^2\hat{\mathbf{z}}$$

so that

$$\langle \mathbf{F}_r \rangle = \frac{e^4}{3m^2c^4}E_0^2\,\hat{\mathbf{z}} = \frac{8\pi}{3}\left(\frac{e^2}{mc^2}\right)^2\frac{E_0^2}{8\pi}\hat{\mathbf{z}} \qquad (S.3.55.4)$$

The radiation reaction force varies with the fourth power of the charge, so a positron would yield the same result.

c) The average power $\langle P \rangle$ scattered by the charge is

$$\langle P \rangle = |\langle \mathbf{S} \rangle|\,\sigma \qquad (S.3.55.5)$$

where σ is the total cross section. The average power is then

$$\langle P \rangle = \frac{cE_0^2}{8\pi}\sigma \qquad (S.3.55.6)$$

The average incident momentum per unit time $\mathbf{p}_i$ is given by

$$\left\langle \frac{d\mathbf{p}_i}{dt} \right\rangle = \frac{1}{c}\left\langle \frac{d\mathcal{E}}{dt} \right\rangle \hat{\mathbf{z}} = \frac{E_0^2}{8\pi}\sigma\hat{\mathbf{z}} \qquad (S.3.55.7)$$

where we used the relation $\mathcal{E} = pc$ for radiation. Using the Thomson cross section for σ in (S.3.55.7) gives the reaction force

$$\langle \mathbf{F}_r \rangle = \left\langle \frac{d\mathbf{p_i}}{dt} \right\rangle = \frac{8\pi}{3} \left(\frac{e^2}{mc^2} \right)^2 \frac{E_0^2}{8\pi} \, \hat{\mathbf{z}}$$

This is the same result as in (S.3.55.4).

3.56   Radiation of Accelerating Positron (Princeton, Colorado)

In first approximation, we disregard the radiation loss, i.e., we consider the energy to be constant at any given moment:

$$E = \frac{mv_1^2}{2} = \frac{mv^2}{2} + \frac{Ze^2}{r} \tag{S.3.56.1}$$

From this equation, we may find v as a function of $r(t)$ and then calculate

$$\frac{dE}{dt} = \frac{2e^2 w^2}{3c^3} = \frac{2e^2 \dot{v}^2}{3c^3} \tag{S.3.56.2}$$

where w is the acceleration of the positron. We should check at the end that the energy change due to radiation is small compared to the initial energy. From (S.3.56.1)

$$v = \sqrt{v_1^2 - \frac{2Ze^2}{mr}} \tag{S.3.56.3}$$

$$\dot{v} = \frac{dv}{dr} v = \frac{Ze^2}{mr^2} \tag{S.3.56.4}$$

Substituting (S.3.56.4) into (S.3.56.2) and integrating, we have

$$\Delta E = \int \frac{2e^2}{3c^3} \left(\frac{Ze^2}{mr^2} \right)^2 dt$$

$$\tag{S.3.56.5}$$

$$= \frac{2e^6 Z^2}{3c^3 m^2} \int \frac{dr}{v_1 r^4 \sqrt{1 - \frac{2Ze^2}{mv_1^2} \frac{1}{r}}}$$

We should integrate (S.3.56.5) from ∞ to $r_{\min}$ and then from $r_{\min}$ to ∞, again only when $\Delta E \ll E_0$. In our approximation, we can say that the radiation during the deceleration is the same as for the period of acceleration and simply write

$$\Delta E = 2 \cdot \frac{2e^6 Z^2}{3c^3 m^2 v_1} \int_{r_{min}}^{\infty} \frac{dr/r^4}{\sqrt{1 - r_{min}/r}} \qquad (S.3.56.6)$$

where $r_{min} = \dfrac{2Ze^2}{mv_1^2}$. Substituting $\eta \equiv r_{min}/r$ (S.3.56.6) becomes

$$\Delta E = \frac{4}{3} \frac{e^6 Z^2}{c^3 m^2 v_1} \left(\frac{1}{r_{min}}\right)^3 \int_0^1 \frac{\eta^2 \, d\eta}{\sqrt{1 - \eta}} = \frac{1}{6} \frac{mv_1^5}{c^3 Z} \int_0^1 \frac{\eta^2 d\eta}{\sqrt{1 - \eta}} \qquad (S.3.56.7)$$

Integrating by parts,

$$\int_0^1 \frac{\eta^2 d\eta}{\sqrt{1 - \eta}} = -\left. 2\eta^2 \sqrt{1 - \eta} \right|_0^1 + 4 \int_0^1 \eta \sqrt{1 - \eta}$$

$$= -\left. \frac{8}{3}\eta(1 - \eta)^{3/2} \right|_0^1 - \frac{8}{3} \cdot \frac{2}{5}(1 - \eta)^{5/2} \Big|_0^1 = \frac{16}{15}$$

Therefore,

$$\Delta E = \frac{16}{15} \cdot \frac{1}{6} \frac{mv_1^5}{c^3 Z} = \frac{8}{45} \frac{mv_1^5}{c^3 Z}$$

Check our initial assumption:

$$\frac{\Delta E}{E} \approx \left(\frac{v_1}{c}\right)^3 \frac{1}{Z} \ll 1$$

So

$$\frac{mv_2^2}{2} = \frac{mv_1^2}{2} - \Delta E = \frac{mv_1^2}{2} - \frac{8}{45} \frac{mv_1^5}{c^3 Z}$$

$$v_2 \approx \left(v_1^2 - \frac{16}{45} \frac{v_1^5}{c^3 Z}\right)^{1/2} \approx v_1 \left(1 - \frac{8}{45} \frac{v_1^3}{c^3 Z}\right)$$

3.57 Half-Wave Antenna (Boston)

a) The vector potential may be found from the integral (see, for instance, Marion and Heald, *Classical Electromagnetic Radiation*, Chapter 8):

$$\mathbf{A}(\mathbf{x}, t) = \frac{1}{c} \int_V dv' \int dt' \frac{\mathbf{J}(\mathbf{x}', t')}{|\mathbf{x} - \mathbf{x}'|} \delta\left(t' + \frac{|\mathbf{x} - \mathbf{x}'| - ct}{c}\right) \qquad (S.3.57.1)$$

The current density may be written with a complex time dependence (taking the real part at the end of the calculation):

$$\mathbf{J}(\mathbf{x},t) = I_0\hat{\mathbf{z}}\cos\left(\frac{2\pi z}{\lambda}\right)\delta(x)\delta(y)e^{i\omega t} \qquad (S.3.57.2)$$

Substituting (S.3.57.2) into (S.3.57.1) and integrating over x' and y', we obtain

$$\mathbf{A}(\mathbf{x},t) \approx \frac{I_0\hat{\mathbf{z}}}{cr}\int_{-\lambda/4}^{\lambda/4}dz'\cos\left(\frac{2\pi z'}{\lambda}\right)e^{i\omega\left(t-(1/c)\sqrt{x^2+y^2+(z-z')^2}\right)} \qquad (S.3.57.3)$$

where we have used the assumption that we are in the radiation zone, so that

$$|\mathbf{x}-\mathbf{x}'| \approx r$$

Expanding the square root in (S.3.57.3) to order z'/r, we find

$$\mathbf{A}(\mathbf{x},t) \approx \frac{I_0\hat{\mathbf{z}}}{cr}e^{i(\omega t-kr)}\int_{-\lambda/4}^{\lambda/4}dz'\cos\left(\frac{2\pi z'}{\lambda}\right)e^{i\frac{2\pi}{\lambda}\frac{z}{r}z'} \qquad (S.3.57.4)$$

Letting $z = r\cos\theta$ and performing the integral in (S.3.57.4) (write cos as sum of exponentials), we get

$$\mathbf{A}(\mathbf{x},t) \approx \frac{I_0\hat{\mathbf{z}}}{cr}\cdot\frac{\lambda}{2\pi}\left[\frac{\sin\left(\frac{\pi}{2}(1+\cos\theta)\right)}{1+\cos\theta}+\frac{\sin\left(\frac{\pi}{2}(1-\cos\theta)\right)}{1-\cos\theta}\right]e^{i(\omega t-kr)}$$

$$(S.3.57.5)$$

$$= \frac{I_0\lambda\hat{\mathbf{z}}}{\pi cr}\frac{\cos\left(\frac{\pi}{2}\cos\theta\right)}{\sin^2\theta}e^{i(\omega t-kr)}$$

b) The electric field $\mathbf{E}$ in the radiation zone may be found directly from

$$\mathbf{E} = -\frac{1}{c}\frac{\partial\mathbf{A}}{\partial t} = -i\frac{\omega}{c}\mathbf{A}$$

using (S.3.57.5). The magnetic induction $\mathbf{B}$ in the radiation zone is given by

$$\mathbf{B} = -\frac{1}{c}\mathbf{n}\times\frac{\partial\mathbf{A}}{\partial t}$$

So

$$\mathbf{B} = -ik(\mathbf{n}\times\mathbf{A})$$

c) The power radiated is calculated using the hint in the problem:

$$\frac{dP}{d\Omega} = \frac{c}{8\pi}\frac{\omega k}{c}\text{Re}\left[r^2\mathbf{n}\cdot(\mathbf{A}\times(\mathbf{n}\times\mathbf{A}))\right]$$

$$= \frac{c}{8\pi}\frac{\omega k}{c}\frac{I_0^2\lambda^2}{\pi^2 c^2 r^2}r^2\frac{\cos^2\left(\frac{\pi}{2}\cos\theta\right)}{\sin^4\theta}\mathbf{n}\cdot(\mathbf{e}_z\times(\mathbf{n}\times\mathbf{e}_z))$$

$$= \frac{I_0^2}{2\pi c}\frac{\cos^2\left(\frac{\pi}{2}\cos\theta\right)}{\sin^4\theta}\mathbf{n}\cdot(\mathbf{n}-\mathbf{e}_z(\mathbf{n}\cdot\mathbf{e}_z)) = \frac{I_0^2}{2\pi c}\frac{\cos^2\left(\frac{\pi}{2}\cos\theta\right)}{\sin^4\theta}\left(1-\cos^2\theta\right)$$

$$= \frac{I_0^2}{2\pi c}\frac{\cos^2\left(\frac{\pi}{2}\cos\theta\right)}{\sin^2\theta}$$

3.58   Čerenkov Radiation (Stony Brook)

a) At each point in the passage of the charged particle through the medium, a spherical wave is produced whose rate of travel is c/n, while the particle is travelling at a velocity v (see Figure S.3.58a). The lines perpendicular to the wavefront give the direction of propagation of the radiation and the angle θ_c, as required:

$$\cos\theta_c = \frac{1}{n\beta} \qquad (\beta \equiv \frac{v}{c})$$

b) The spherical mirror and the cone of radiation produced by the charged particle are azimuthally symmetric, so we may solve the problem in two dimensions. We now must show that the parallel rays striking the mirror

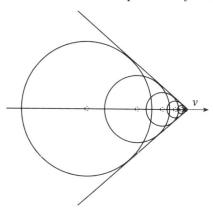

Figure   S.3.58a

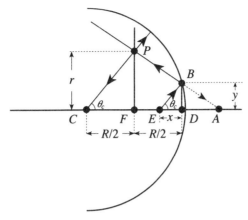

Figure   S.3.58b

will be focussed to a point on the "focal line" of the mirror. The focal
length of a spherical mirror with radius of curvature R is $R/2$. Consider
two rays incident on the mirror at an angle θ_c to the horizontal, one which
passes through the center of the circle and another on the opposite side of
the focus, which strikes the mirror a a distance x away as measured along
the axis (see Figure S.3.58b). In the paraxial approximation, we may use
the standard relation between the image distance i, the object distance o,
and the focal length $f = R/2$

$$\frac{1}{i} + \frac{1}{o} = \frac{1}{f} \qquad (S.3.58.1)$$

Checking for the ray leaving the center of the circle (object at 2f), we have

$$\frac{1}{i} + \frac{1}{R} = \frac{2}{R}$$

$$i = R$$

Obviously, a ray along a radius of the circle will be reflected back on itself.
It passes through the point P along the focal line at a distance $(R/2)\tan\theta_c$
from the axis. Now, a ray originating at the point E will strike the mirror
at the point B and form a virtual image at point A on the other side of the
mirror. Using (S.3.58.1) to find the image distance i and thereby $\overline{AD}$,

$$\frac{1}{i} + \frac{1}{x} = \frac{2}{R} \qquad (S.3.58.2)$$

$$\overline{AD} = |i| = \frac{Rx}{R - 2x} \qquad (S.3.58.3)$$

where the length of the segment $\overline{AD}$ is made positive (although the image distance is negative). To find the point where the reflected ray crosses the focal line, we will use similar triangles, $\triangle APF \sim \triangle ABD$. So

$$\frac{\overline{PF}}{\overline{BD}} = \frac{\overline{AF}}{\overline{AD}} \qquad (S.3.58.4)$$

First, we find that $y = \overline{BD} = x \tan \theta_c$, and then in the same paraxial approximation (see Figure S.3.58b)

$$\frac{r}{y} = \frac{(Rx)/(R - 2x) + R/2}{(Rx)/(R - 2x)} \qquad (S.3.58.5)$$

so

$$r = \frac{Rx + R^2/2 - Rx}{Rx} y = \left(\frac{R}{2x}\right) x \tan \theta_c = \frac{R}{2} \tan \theta_c \quad (S.3.58.6)$$

c) Restoring the cylindrical symmetry to the problem, the point in the focal line becomes a circle in the focal plane of radius

$$r = \frac{R}{2} \tan \theta_c$$

3.59   Stability of Plasma (Boston)

a) Calculate the force on the ion of charge Q at radius R in cylindrical coordinates with unit vectors $\hat{\mathbf{r}}$, $\mathbf{e}_\varphi$, and $\hat{\mathbf{z}}$. The electrostatic force

$$\mathbf{F}_e = \frac{2\lambda Q}{R} \hat{\mathbf{r}} \qquad (S.3.59.1)$$

where $\lambda = I/v$ is the linear charge density. So

$$\mathbf{F}_e = \frac{2IQ}{vR} \hat{\mathbf{r}}$$

and the magnetic force

$$\mathbf{F}_m = \frac{Q}{c} \mathbf{v} \times \mathbf{B} = \frac{Q}{c} \frac{2I}{cR} \mathbf{v} \times \mathbf{e}_\varphi \qquad (S.3.59.2)$$

where $\mathbf{B} = \dfrac{2I}{cR} \mathbf{e}_\varphi$ could be found from

$$\oint \mathbf{B} \cdot d\mathbf{l} = 2\pi RB = \int_S \frac{4\pi}{c} \mathbf{J} \cdot d\mathbf{S} = \frac{4\pi}{c} I \qquad (S.3.59.3)$$

So
$$\mathbf{F_m} = -\frac{2IQv}{c^2R}\,\hat{\mathbf{r}} \qquad (S.3.59.4)$$

Therefore, the total force acts outward
$$\mathbf{F} = \mathbf{F_e} + \mathbf{F_m} = \frac{2IQ}{Rv}\,\hat{\mathbf{r}} - \frac{2IQv}{c^2R}\,\hat{\mathbf{r}} = \frac{2IQ}{Rv}\left(1 - \frac{v^2}{c^2}\right)\hat{\mathbf{r}} \qquad (S.3.59.5)$$

b) The force calculated in (a) does some work on an ion, which leads to an increase of its kinetic energy (we take the initial radial velocity of the ion to be zero)
$$\int_R^r \mathbf{F(r)} \cdot d\mathbf{r} = \frac{1}{2}M\left(\frac{dr}{dt}\right)^2 \qquad (S.3.59.6)$$

Performing the integration in (S.3.59.6), we obtain
$$\frac{2IQ}{v}\left(1 - \frac{v^2}{c^2}\right)\ln\frac{r}{R} = \frac{1}{2}M\left(\frac{dr}{dt}\right)^2$$

or
$$\frac{dr}{dt} = \sqrt{\frac{4IQ}{Mv}\left(1 - \frac{v^2}{c^2}\right)\ln\frac{r}{R}}$$

3.60 Charged Particle in Uniform Magnetic Field (Princeton)

The solution to this problem is similar to that for Problem P.1.52, where we considered the adiabatic invariant for a mechanical system. Here, we have for the motion in the plane perpendicular to the magnetic field
$$I = \frac{1}{2\pi}\oint \mathbf{P_t} \cdot d\mathbf{r} = \text{const}$$

where
$$\mathbf{P_t} = \mathbf{p_t} + \frac{e}{c}\mathbf{A}$$

is the projection of the generalized momentum on this plane, and the integral is taken over one period of motion in this plane, whose shape is a circle. ($\mathbf{A}$ is the vector potential and e is the charge of the particle.)
$$I = \frac{1}{2\pi}\oint \mathbf{p_t} \cdot d\mathbf{r} + \frac{e}{2\pi c}\oint \mathbf{A} \cdot d\mathbf{r} \qquad (S.3.60.1)$$

Using Stokes' theorem and substituting $\mathbf{H} = \nabla \times \mathbf{A}$, we obtain
$$I = p_t r + \frac{e}{2\pi c}\int_S \mathbf{H} \cdot d\mathbf{S} = p_t r - \frac{eH}{2c}r^2 \qquad (S.3.60.2)$$

We used the fact that the absolute value of $\mathbf{p}_t$ is constant. The minus sign before the second term occurs since the line integral about the orbit is opposite to the velocity of the charge. After substituting $p_t = eHr/c$ into (S.3.60.2) (see Problem 3.63), we obtain

$$I = \frac{eH}{c}r^2 - \frac{eH}{2c}r^2 = \frac{eH}{2c}r^2 = \text{const} \qquad \text{(S.3.60.3)}$$

So, for a slow change of magnetic field from H_0 to H_1, we find

$$H_0 R_0^2 = H_1 R_1^2$$

or

$$R_1 = R_0\sqrt{\frac{H_0}{H_1}}$$

Now, if the field changes suddenly from H_1 back to H_0, then the energy $\mathcal{E}$ is conserved

$$\mathcal{E} = \frac{1}{2}m\omega^2 R^2$$

$$\omega_1 R_1 = \omega_0 R_2$$

where ω_1 and ω_0 are cyclotron frequencies, corresponding to magnetic fields H_1 and H_0, respectively:

$$\omega = \frac{eH}{mc}$$

where m is the mass of the particle. Therefore,

$$R_2 = R_1\frac{\omega_1}{\omega_0} = R_0\frac{H_1}{H_0}\sqrt{\frac{H_0}{H_1}} = R_0\sqrt{\frac{H_1}{H_0}}$$

3.61 Lowest Mode of Rectangular Wave Guide (Princeton, MIT, Michigan State)

a) Because the walls are perfectly conducting, we have for $\mathbf{E}$ and $\mathbf{B}$ the boundary conditions

$$\mathbf{n} \times \mathbf{E} = 0$$

$$\mathbf{n} \cdot \mathbf{B} = 0$$

where $\mathbf{n}$ is normal to the wall, or in terms of $E_z(B_z)$ (z is the direction of wave propagation)

$$E_z|_S = 0 \qquad \frac{\partial B_z}{\partial n}\bigg|_S = 0 \qquad \text{(S.3.61.1)}$$

b) Starting from the sourceless Maxwell equations in vacuum

$$\nabla \times \mathbf{E} = -\frac{1}{c}\frac{\partial \mathbf{B}}{\partial t} \qquad (S.3.61.2)$$

$$\nabla \times \mathbf{B} = \frac{1}{c}\frac{\partial \mathbf{E}}{\partial t} \qquad (S.3.61.3)$$

$$\nabla \cdot \mathbf{B} = 0 \qquad (S.3.61.4)$$

$$\nabla \cdot \mathbf{E} = 0 \qquad (S.3.61.5)$$

and substituting $\mathbf{E}(\mathbf{r},t) = \mathbf{E}(\mathbf{r})\exp(-i\omega t)$ (same for $\mathbf{B}$), we obtain

$$\nabla \times \mathbf{E} = \frac{i\omega}{c}\mathbf{B} \qquad (S.3.61.6)$$

$$\nabla \times \mathbf{B} = -\frac{i\omega}{c}\mathbf{E} \qquad (S.3.61.7)$$

$$\nabla \cdot \mathbf{B} = 0 \qquad (S.3.61.8)$$

$$\nabla \cdot \mathbf{E} = 0 \qquad (S.3.61.9)$$

The field dependence on z may be written in the form $f(z) = \exp\left[i(kz - \omega t)\right]$, where k is the wave vector for the wave transmitted in the z direction. Using the fact that the electric field of the lowest mode is in the y direction only, we have, from (S.3.61.6)–(S.3.61.7),

$$-ikE_y = \frac{i\omega}{c}B_x \qquad (S.3.61.10)$$

$$0 = B_y \qquad (S.3.61.11)$$

$$\frac{\partial B_z}{\partial y} = 0 \qquad (S.3.61.12)$$

$$ikB_x - \frac{\partial B_z}{\partial x} = -\frac{i\omega}{c}E_y \qquad (S.3.61.13)$$

From (S.3.61.12), $B_z = B(x)\exp(ikz)$, and substituting (S.3.61.10) into

(S.3.61.13), we obtain

$$B_x = \frac{ik}{\gamma^2} \frac{\partial B_z}{\partial x} \tag{S.3.61.14}$$

$$E_y = -\frac{i\omega}{c\gamma^2} \frac{\partial B_z}{\partial x} \tag{S.3.61.15}$$

where

$$\gamma^2 \equiv \frac{\omega^2}{c^2} - k^2 \tag{S.3.61.16}$$

Using $\nabla \cdot \mathbf{B} = 0$ and B_x from (S.3.61.14), we get a differential equation for B_z

$$\frac{ik}{\gamma^2} \frac{\partial^2 B_z(x)}{\partial x^2} + ikB_z(x) = 0 \tag{S.3.61.17}$$

or

$$\frac{\partial^2 B_z(x)}{\partial x^2} + \gamma^2 B_z(x) = 0 \tag{S.3.61.18}$$

The solution of this equation satisfying the boundary conditions

$$\left. \frac{\partial B_z}{\partial n} \right|_S = 0$$

is $B_z = B_0 \cos \gamma x$ with $\gamma = \pi/a$. So the field in the wave guide in this mode from (S.3.61.14)– (S.3.61.15)

$$B_z = B_0 \cos \left(\frac{\pi}{a} x \right) \cdot e^{ikz - i\omega t} \tag{S.3.61.19}$$

$$B_x = -\frac{ika}{\pi} B_0 \sin \left(\frac{\pi}{a} x \right) \cdot e^{ikz - i\omega t} \tag{S.3.61.20}$$

$$E_y = \frac{i\omega a}{c\pi} B_0 \sin \left(\frac{\pi}{a} x \right) \cdot e^{ikz - i\omega t} \tag{S.3.61.21}$$

c) The dispersion relation for the lowest mode is found from (S.3.61.16):

$$\omega = c\sqrt{k^2 + \pi^2/a^2} \tag{S.3.61.22}$$

The phase velocity v is

$$v = \frac{\omega}{k} = c\sqrt{1 + \pi^2/(k^2 a^2)} \tag{S.3.61.23}$$

The group velocity u is

$$u = \frac{\partial \omega}{\partial k} = \frac{ck}{\sqrt{k^2 + \pi^2/a^2}} \tag{S.3.61.24}$$

d) The waves propagating in the wave guides can be divided into two classes: TE (transverse electric, $E_z = 0$) as is the case in this problem, and TM (transverse magnetic, $B_z = 0$).

3.62 TM Modes in Rectangular Wave Guide (Princeton)

a) Again, as in Problem 3.61, we can express all the fields in terms of a single longitudinal component. In this problem, we are considering TM waves so $B_z = 0$, and we use E_z instead. We find for the field components

$$E_x = \frac{ik}{\gamma^2} \frac{\partial E_z}{\partial x} \qquad\qquad E_y = \frac{ik}{\gamma^2} \frac{\partial E_z}{\partial y}$$

$$\tag{S.3.62.1}$$

$$B_x = -\frac{i\omega}{c\gamma^2} \frac{\partial E_z}{\partial y} \qquad\qquad B_y = \frac{i\omega}{c\gamma^2} \frac{\partial E_z}{\partial x}$$

where again

$$\gamma^2 = \frac{\omega^2}{c^2} - k^2$$

The wave equation for the E_z component is

$$\nabla_t^2 E_z + \gamma^2 E_z = 0 \tag{S.3.62.2}$$

where

$$\nabla_t^2 \equiv \frac{\partial^2}{\partial x^2} + \frac{\partial^2}{\partial y^2}$$

The solution to (S.3.62.2) with the boundary condition $\mathbf{E} \times \mathbf{n}|_S = 0$ is given by

$$E_z = E_0 \sin \frac{\pi n}{a} x \cdot \sin \frac{\pi m}{b} y \tag{S.3.62.3}$$

where n and m are integers. So

$$\gamma^2 = \pi^2 \left(\frac{n^2}{a^2} + \frac{m^2}{b^2} \right) \tag{S.3.62.4}$$

The frequency ω is given by

$$\omega = c\sqrt{k^2 + \gamma^2} = c\sqrt{k^2 + \pi^2 \left(\frac{n^2}{a^2} + \frac{m^2}{b^2} \right)} \qquad \text{(S.3.62.5)}$$

The cutoff frequency corresponds to $k = 0$, so

$$\omega = c\pi\sqrt{\frac{n^2}{a^2} + \frac{m^2}{b^2}} \qquad \text{(S.3.62.6)}$$

For TM waves, we cannot take any of the n or $m = 0$ modes because that would make $E_z = 0$. So the lowest cutoff frequency corresponds to $m = n = 1$

$$\omega_{11} = c\pi\sqrt{\frac{1}{a^2} + \frac{1}{b^2}} = 3 \cdot 10^{10} \cdot 3.14\sqrt{\frac{1}{(7.21)^2} + \frac{1}{(3.4)^2}} \approx 3.1 \cdot 10^{10} \text{s}^{-1}$$

So the TM radiation with frequency $6.1 \cdot 10^{10} \text{s}^{-1}$ will propagate in the guide.

b) The dispersion relation was given in (a):

$$\omega = c\sqrt{k^2 + \pi^2 \left(\frac{n^2}{a^2} + \frac{m^2}{b^2} \right)}$$

c) The wave number as a function of the cutoff frequency ω_λ can be written in the form

$$k_\lambda = \frac{1}{c}\sqrt{\omega^2 - \omega_\lambda^2}$$

The wave of frequency $\omega < \omega_\lambda$ cannot propagate (k becomes imaginary), and in fact the attenuation of the field will be given by k_λ. In our case,

$$k_\lambda = \frac{1}{c}\sqrt{\left(\frac{\omega_\lambda}{2} \right)^2 - \omega_\lambda^2} = \frac{\sqrt{3}}{2}\frac{\omega_\lambda}{c} i$$

We may write E_z in the form

$$E_z \propto E_0 e^{ik_\lambda z} = E_0 e^{-(\sqrt{3}\,\omega_\lambda z)/(2c)}$$

The power dissipation will be proportional to $|E_z|^2$:

$$P(z) \propto e^{-(\sqrt{3}\,\omega_\lambda z)/c}$$

We wish to find the point where $P(z)/P(0) = 1/e$. Hence,

$$z = \frac{c}{\omega_\lambda \sqrt{3}} = \frac{3 \cdot 10^{10}}{3.1 \cdot 10^{10} \sqrt{3}} = 0.56 \text{ cm}$$

3.63 Betatron (Princeton, Moscow Phys-Tech, Colorado, Stony Brook (a))

a) Assume we have a magnetic field that is constant along and perpendicular to the plane of the orbit $\mathbf{B}(z = 0) = B(R)\hat{\mathbf{z}}$ (see Figure S.3.63). The Lorentz force gives

$$\dot{\mathbf{p}} = \frac{e}{c}\mathbf{v} \times \mathbf{B} \tag{S.3.63.1}$$

We can substitute the energy $\mathcal{E}$ for the momentum by using

$$\mathbf{p} = \frac{\mathcal{E}\mathbf{v}}{c^2}$$

and since the energy does not change in the magnetic field we have

$$\frac{\mathcal{E}}{c^2}\frac{d\mathbf{v}}{dt} = \frac{e}{c}\mathbf{v} \times \mathbf{B} \tag{S.3.63.2}$$

or, separating into components,

$$\dot{v}_x = \Omega v_y \tag{S.3.63.3}$$

$$\dot{v}_y = -\Omega v_x \tag{S.3.63.4}$$

$$\dot{v}_z = 0$$

where

$$\Omega = \frac{ecB(R)}{\mathcal{E}} \tag{S.3.63.5}$$

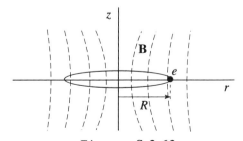

Figure S.3.63

Following a standard procedure, we multiply (S.3.63.4) by $i = \sqrt{-1}$ and add it to (S.3.63.3), which yields

$$\dot{u} + i\Omega u = 0 \qquad (S.3.63.6)$$

where $u = v_x + iv_y$ or

$$u = Ae^{-i(\Omega t + \alpha)} \qquad (S.3.63.7)$$

where A and α are real. Separating real and imaginary parts of (S.3.63.7), we obtain

$$v_x = A\cos(\Omega t + \alpha) \qquad (S.3.63.8)$$

$$v_y = -A\sin(\Omega t + \alpha) \qquad (S.3.63.9)$$

From (S.3.63.8) and (S.3.63.9), we can see that

$$v_x^2 + v_y^2 = A^2 = v_0^2$$

where v_0 is the initial velocity of the particle, which as we assumed moves only in the x–y plane. Integrating again, we find

$$x = x_0 + \frac{v_0}{\Omega}\sin(\Omega t + \alpha)$$
$$\qquad (S.3.63.10)$$
$$y = y_0 + \frac{v_0}{\Omega}\cos(\Omega t + \alpha)$$

So the radius R is given by

$$R^2 = (x - x_0)^2 + (y - y_0)^2 = \frac{v_0^2}{\Omega^2} \qquad (S.3.63.11)$$

and

$$R = \frac{v_0}{|\Omega|} = \frac{v_0 \mathcal{E}}{ecB(R)} = \frac{cp_0}{eB(R)} \qquad (S.3.63.12)$$

b) From (S.3.63.12), the momentum

$$p = \frac{eB_0 R}{c} \qquad (S.3.63.13)$$

where $B_0 = B_0(R)$. Assuming that R does not change, we find from (S.3.63.13) that

$$\Delta p = \frac{eR}{c}\Delta B_0 \qquad (S.3.63.14)$$

If the magnetic field through the orbit is increased, a tangential electric field will be produced at the position of the orbit:

$$E_\varphi = \frac{\dot{\Phi}}{2\pi Rc} = \frac{\langle \dot{B} \rangle \pi R^2}{2\pi Rc} = \frac{\langle \dot{B} \rangle R}{2c} \qquad \text{(S.3.63.15)}$$

Therefore the rate of increase of the momentum is

$$\dot{p} = eE_\varphi = \frac{eR\langle \dot{B} \rangle}{2c} \qquad \text{(S.3.63.16)}$$

Integrating (S.3.63.16), we obtain

$$\Delta p = \frac{eR\langle \Delta B \rangle}{2c} \qquad \text{(S.3.63.17)}$$

Equating (S.3.63.14) and (S.3.63.17), we have

$$\langle \Delta B \rangle = 2\Delta B_0 \qquad \text{(S.3.63.18)}$$

indicating that the change in flux through the orbit must be twice that which would have been obtained if the magnetic field were spatially uniform (Betatron rule 2:1).

c) Consider first the vertical displacement (we assume that the vertical and radial motions are decoupled)

$$\frac{dp_z}{dt} = \frac{d(\gamma m v_z)}{dt} = F_z \qquad \text{(S.3.63.19)}$$

where $\gamma = 1/\sqrt{1 - v^2/c^2}$. Since v_z is much smaller than the velocity in the x–y plane, we disregard any change in γ due to changes in v_z

$$\frac{d(\gamma m v_z)}{dt} = \gamma m \frac{dv_z}{dt} = \gamma m \ddot{z} = \frac{e}{c} v B_r(R) \qquad \text{(S.3.63.20)}$$

where $B_r(R)$ is the radial **B** field at a radius R. Neglecting the space charge current and displacement current and using cylindrical coordinates, we may write

$$(\nabla \times \mathbf{B})_\varphi = 0 = \left(\frac{\partial B_r}{\partial z} - \frac{\partial B_z}{\partial r} \right)$$

So

$$\frac{\partial B_r}{\partial z} = \frac{\partial B_z}{\partial r}$$

or, for small z,

$$B_r(R) = \left(\frac{\partial B_r}{\partial z}\right)_{r=R} z = \left(\frac{\partial B_z}{\partial r}\right)_{r=R} z \qquad \text{(S.3.63.21)}$$

Using the expression for B_z given in the problem,

$$B_z(r) = B_0 \left(\frac{R}{r}\right)^n$$

and substituting it into (S.3.63.21) and then (S.3.63.20), we obtain

$$\gamma m \ddot{z} = -n\frac{eB_0 v}{cR}\left(\frac{R}{r}\right)^{n+1}\Bigg|_{r=R} z = -\frac{neB_0 v}{cR} z \qquad \text{(S.3.63.22)}$$

So

$$\ddot{z} = -\frac{neB_0 v}{\gamma cmR} z \qquad \text{(S.3.63.23)}$$

Taking $v = \Omega R$ from (S.3.63.12), (S.3.63.23) becomes

$$\ddot{z} = -n\frac{eB_0}{\gamma mc}\Omega z = -n\frac{ecB_0}{\mathcal{E}}\Omega z = -n\Omega^2 z \qquad \text{(S.3.63.24)}$$

Therefore (S.3.63.24) exhibits oscillatory behavior along the z-axis which is stable if $n\Omega^2 > 0$ and so $n > 0$. The period of oscillation is then

$$\omega_z = \Omega\sqrt{n} \qquad \text{(S.3.63.25)}$$

For the radial motion, the Lorentz force F_L is

$$F_L = -\frac{e}{c}v B_z(r) \qquad \text{(S.3.63.26)}$$

For small deviations from equilibrium $\rho \equiv r - R$, we can write (S.3.63.26) in the form

$$F_L = -\frac{e}{c}vB_0(R) - \frac{e}{c}v\frac{\partial B}{\partial r}\bigg|_{r=R}(r - R) = F_L(R) + \frac{e}{c}v\frac{n}{R}B_0\rho$$

$$= F_L(R) + \frac{eB_0\Omega n}{c}\rho = F_L(R) + \gamma m\Omega^2 n\rho \qquad \text{(S.3.63.27)}$$

where we again used (S.3.63.12) for the cyclotron frequency. We must also consider the centrifugal force

$$F_c = \gamma m\omega^2 r = \gamma m\frac{\Omega^2 R^2}{r} \qquad \text{(S.3.63.28)}$$

where we used the conservation of canonical angular momentum

$$\mathcal{P}_\varphi = \gamma m r \dot\varphi = \gamma m r \omega$$

Now, for small ρ,

$$F_c = \gamma m \frac{\Omega^2 R^2}{R + \rho} = \gamma m \Omega^2 R \left(1 - \frac{\rho}{R}\right) = F_c(R) - \gamma m \Omega^2 \rho \qquad \text{(S.3.63.29)}$$

Combining (S.3.63.27) and (S.3.63.29), we obtain

$$\gamma m \ddot\rho = F_c(R) + F_L(R) + \gamma m \Omega^2 \rho n - \gamma m \Omega^2 \rho \qquad \text{(S.3.63.30)}$$

Since $F_c(R) + F_L(R) = 0$ at equilibrium, we can write

$$\ddot\rho = (n - 1)\Omega^2 \rho \qquad \text{(S.3.63.31)}$$

Again, as for the vertical motion, we have an oscillation of frequency

$$\omega_r = \Omega\sqrt{1 - n}$$

This oscillation is stable if $1 - n > 0$, or $n < 1$.

d) The condition for both radial and vertical stability will be the intersection of the two conditions for n, so

$$0 < n < 1$$

A more detailed discussion of this problem may be found in *Phys. Rev.* **60**, 53 (1941).

3.64   Superconducting Frame in Magnetic Field (Moscow Phys-Tech)

Find the magnetic field from the vector potential

$$\mathbf{B} = \nabla \times \mathbf{A} = \begin{vmatrix} \hat{\mathbf{x}} & \hat{\mathbf{y}} & \hat{\mathbf{z}} \\ \dfrac{\partial}{\partial x} & \dfrac{\partial}{\partial y} & \dfrac{\partial}{\partial z} \\ -B_0 y & \alpha x z & 0 \end{vmatrix} = (-\alpha x, 0, \alpha z + B_0)$$

The magnetic flux Φ through the surface of the superconducting frame is constant (see Figure S.3.64). Φ is composed of the flux from the external

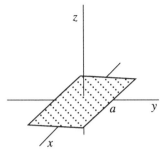

Figure S.3.64

magnetic field Φ_e and the flux Φ_i produced by the current I flowing in the frame:

$$\Phi = \Phi_e + \Phi_i = \mathbf{B} \cdot \mathbf{S} + \frac{1}{c}LI = (B_0 + \alpha z)a^2 + \frac{1}{c}LI$$

At $t = 0$, $z = 0$ and $\Phi = \Phi_0 = B_0 a^2$. At later times,

$$\Phi = B_0 a^2 + \alpha z a^2 + \frac{1}{c}LI = \Phi_0$$

So for the current we have

$$I = -\frac{c\alpha z a^2}{L}$$

The force on the frame due to the interaction with the magnetic field is given by the general formula

$$\mathbf{F} = \frac{I}{c} \oint d\mathbf{l} \times \mathbf{B}$$

In this problem, due to the physical constraint, we need only the component in the z direction

$$F_z = \frac{2I}{c} \int\limits_{-a/2}^{a/2} (-dy)(-\alpha a/2) = \frac{2I}{c}\frac{\alpha a^2}{2} = \frac{\alpha I}{c}a^2 = -\frac{\alpha^2 a^4}{L}z$$

Therefore, the equation of motion becomes

$$m\ddot{z} = -\frac{\alpha^2 a^4}{L}z - mg$$

or

$$\ddot{z} + \frac{\alpha^2 a^4}{mL}z = -g$$

This equation describes oscillatory motion with $\omega = \alpha a^2/\sqrt{mL}$, and the solution is
$$z = A\cos(\omega t + \varphi_0) + z_0$$
where
$$z_0 = -\frac{mgL}{\alpha^2 a^4} = -\frac{g}{\omega^2}$$
From the initial conditions $z(0) = 0$ and $\dot{z}(0) = 0$, we find that $A = -z_0$ and $\varphi_0 = 0$. The complete solution for the position of the frame along the z-axis at time t is
$$z = z_0(1 - \cos\omega t) = \frac{g}{\omega^2}(\cos\omega t - 1)$$

3.65 Superconducting Sphere in Magnetic Field (Michigan State, Moscow Phys-Tech)

a) From symmetry considerations, it is clear that the current would flow on the surface of the shell perpendicular to the applied magnetic field. As for any ellipsoid in a uniform electric or magnetic field (see Problem 3.10), we can assume that the field outside the shell produced by these currents is equivalent to a magnetic dipole moment $\mathbf{m}$ placed in the center of the shell. For
$$\mathbf{B}_m = \frac{3(\mathbf{m} \cdot \mathbf{r})\mathbf{r}}{r^5} - \frac{\mathbf{m}}{r^3}$$
The total field outside is then $\mathbf{B} = \mathbf{B}_m + \mathbf{B}_0$. The boundary condition on the surface at an arbitrary point gives

$$B_{0n} + B_{mn} = 0$$

The normal component of $\mathbf{B}$ is continuous and inside $\mathbf{B} = 0$. From the boundary conditions on the surface at an arbitrary angle θ between the direction of $\mathbf{B}_0$ and the normal $\mathbf{n}$ (see Figure S.3.65) we have

$$B_{mn} = \frac{3m\cos\theta}{R^3} - \frac{m\cos\theta}{R^3} = \frac{2m\cos\theta}{R^3} \qquad (S.3.65.1)$$

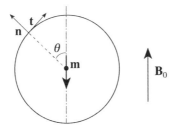

Figure S.3.65

Hence

$$B_0 \cos \theta + \frac{2m \cos \theta}{R^3} = 0 \qquad \text{(S.3.65.2)}$$

At $\mathbf{m} = -\left(R^3/2\right) \mathbf{B}_0$, where R is the radius of the spherical shell, the boundary conditions are satisfied on the surface of the shell. Therefore,

$$\mathbf{B} = \mathbf{B}_m + \mathbf{B}_0 = \mathbf{B}_0 - \frac{3R^3}{2} \frac{(\mathbf{B}_0 \cdot \mathbf{r})\mathbf{r}}{r^5} + \frac{R^3 \mathbf{B}_0}{2r^3} = \left(1 + \frac{R^3}{2r^3}\right) \mathbf{B}_0 - \frac{3R^3(\mathbf{B}_0 \cdot \mathbf{r})\mathbf{r}}{2r^5}$$

b) The surface current density J_s can be found by using tangential $\mathbf{H}$ component continuity:

$$\frac{3}{2}B_0 \sin \theta = H_{0t} + H_t = \frac{4\pi}{c} J_s$$

and therefore $J_s(\theta) = 3cB_0 \sin \theta / 8\pi$. This solution is only true while $\mathbf{B} < 2/3 \, \mathbf{H}_c$, and the whole sphere is superconducting. When $\mathbf{B} > 2/3 \, \mathbf{H}_c$, the field at the equator exceeds $\mathbf{H}_c$, and the sphere goes into an intermediate state.

3.66 London Penetration Depth (Moscow Phys-Tech)

a) Equation (P.3.66.1) can be obtained from a naive classical model with a superconducting electron density n_s. This derivation is not quite rigorous since we assume spatially uniform fields in the penetration depth (see, for instance, M.Tinkham, *Introduction to Superconductivity*). For a unit volume of superconducting electrons in an electric field $\mathbf{E}$,

$$n_s m \frac{d\mathbf{v}_s}{dt} = n_s e \mathbf{E} \qquad \text{(S.3.66.1)}$$

The superconducting current density may be written as $\mathbf{J}_s = n_s e \mathbf{v}_s$. Substituting into (S.3.66.1) gives (P.3.66.1):

$$\mathbf{E} = \frac{d}{dt} (\Lambda \mathbf{J}_s)$$

To derive (P.3.66.2), write the kinetic energy density in the form

$$\mathcal{E}_{\text{kin}} = n_s \frac{m v_s^2}{2} = \frac{m J_s^2}{2 n_s e^2}$$

Using Maxwell's equation

$$\nabla \times \mathbf{h} = \frac{4\pi}{c} \mathbf{J}_s$$

we obtain

$$\mathcal{E}_{\text{kin}} = \frac{\lambda_L^2}{8\pi} (\nabla \times \mathbf{h})^2 \qquad \text{(S.3.66.2)}$$

where

$$\lambda_L^2 = \frac{mc^2}{4\pi n_s e^2} \tag{S.3.66.3}$$

Now we can write the free energy in the form

$$\mathcal{F}_{s,h} = \mathcal{F}_{s,0} + \frac{1}{8\pi} \int_V \left(h^2 + \lambda_L^2 \left(\nabla \times h \right)^2 \right) dV$$

where $\int h^2/(8\pi)dV$ accounts for the energy of the magnetic field in the superconductor. We want to find a function $h(r)$ that will minimize the free energy. Note that a more rigorous approach requires that we minimize the Gibbs' free energy, but the result is the same. Using a standard procedure (see Goldstein, *Classical Mechanics*, Chapter 2) we take a variation of $\mathcal{F}_{s,h}$. We can write

$$0 = \frac{1}{8\pi} \int_V \left[2h\delta h + 2\lambda_L^2 \left(\nabla \times h \right) \cdot \left(\nabla \times \delta h \right) \right] dV \tag{S.3.66.4}$$

The second term in (S.3.66.4) can be transformed using the identity

$$\nabla \cdot (a \times b) = b \cdot \nabla \times a - a \cdot \nabla \times b$$

Now we can rewrite (S.3.66.4) in the form

$$\int_V \left[h + \lambda_L^2 \left(\nabla \times \left(\nabla \times h \right) \right) \right] \cdot \delta h dV - \int_V \nabla \cdot \left[\left(\nabla \times h \right) \times \delta h \right] dV = 0 \tag{S.3.66.5}$$

The second integral in (S.3.66.5) can be transformed into a surface integral using Gauss's theorem

$$\int_V \nabla \cdot \left[\left(\nabla \times h \right) \times \delta h \right] dV = \oint_S \left[\left(\nabla \times h \right) \times \delta h \right] dS = 0$$

since everywhere on the surface $h = H_0$, and so $\delta h = 0$. So, from the first integral in (S.3.66.5), we obtain (P.3.66.2)

$$h + \lambda_L^2 \nabla \times \left(\nabla \times h \right) = 0$$

b) Using (P.3.66.2) and the identity

$$\nabla \times \left(\nabla \times h \right) = -\nabla^2 h$$

we have

$$\nabla^2 h - \frac{1}{\lambda_L^2} h = 0$$

Orient the x direction normal to the boundary and the magnetic field parallel to it, $\mathbf{H} = H_0\hat{\mathbf{z}}$. From the symmetry of the problem, it is obvious that $\mathbf{h} = h(x)\hat{\mathbf{z}}$. Then (P.3.66.2) becomes

$$\frac{d^2h}{dx^2} - \frac{1}{\lambda_L^2}h = 0$$

whose solution is

$$h = ae^{-x/\lambda_L} + be^{x/\lambda_L}$$

Invoking the boundary conditions $h(0) = H_0$ and $h(\infty) = 0$, we arrive at

$$h = H_0 e^{-x/\lambda_L}$$

where λ_L is the London penetration depth introduced in (a) (S.3.66.2). For a typical metal superconductor with one free electron per atom

$$n \approx \frac{N_A\rho_m}{A}$$

where N_A is Avogadro's number, ρ_m is the mass density, and A is the atomic mass in grams per mole. For a typical superconductor, $n \approx 10^{22}-10^{23}\text{cm}^{-3}$. Assuming $n_s = n$ at $T = 0$, we have from (S.3.66.3)

$$\lambda_L(0) = \sqrt{\frac{mc^2}{4\pi ne^2}} \approx 200-500\text{Å}$$

3.67 Thin Superconducting Plate in Magnetic Field (Stony Brook)

a) Choose $x = 0$ at the center of the plate (see Figure S.3.67a). We know (see, for instance, Problem 3.66) that the external field penetrates to a depth λ_L into the superconductor; this can be described in our case by the

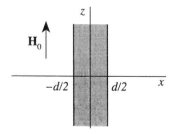

Figure S.3.67a

equation:

$$\nabla_2 \mathbf{H} - \frac{1}{\lambda_L^2} \mathbf{H} = 0$$

Because of the symmetry of the problem, $\mathbf{H} = H(x)\hat{\mathbf{z}}$ where the H field inside the the superconductor will be in the z direction and depend only on x. So we have

$$\frac{d^2 H}{dx^2} - \frac{1}{\lambda_L^2} H = 0 \qquad (S.3.67.1)$$

The general solution of (S.3.67.1) is

$$H = H_1 \cosh(x/\lambda_L) + H_2 \sinh(x/\lambda_L)$$

Using the boundary conditions

$$H\left(\pm \frac{d}{2}\right) = H_0$$

we obtain

$$H_1 = \frac{H_0}{\cosh(d/2\lambda_L)} \qquad H_2 = 0$$

So

$$H = H_0 \frac{\cosh(x/\lambda_L)}{\cosh(d/2\lambda_L)}$$

The supercurrent density J_s can be found from Maxwell's equation

$$\nabla \times \mathbf{H} = \frac{4\pi}{c} \mathbf{J}_s$$

Since $\mathbf{H} = H\hat{\mathbf{z}}$, we have

$$\nabla \times \mathbf{H} = \begin{vmatrix} \hat{\mathbf{x}} & \hat{\mathbf{y}} & \hat{\mathbf{z}} \\ \dfrac{\partial}{\partial x} & \dfrac{\partial}{\partial y} & \dfrac{\partial}{\partial z} \\ 0 & 0 & H \end{vmatrix} = -\frac{\partial H}{\partial x}\hat{\mathbf{y}}$$

and

$$\mathbf{J}_s = J_s(x)\hat{\mathbf{y}} = \frac{c}{4\pi}\left(-\frac{\partial H}{\partial x}\right)\hat{\mathbf{y}} = -\frac{cH_0}{4\pi\lambda_L}\frac{\sinh(x/\lambda_L)}{\cosh(d/2\lambda_L)}\hat{\mathbf{y}}$$

b) In the limiting case of a thin film, $\lambda_L \gg d$, we have

$$H \approx H_0$$

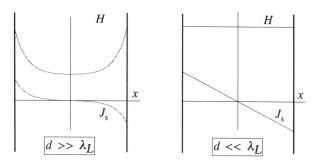

Figure **S.3.67b**

$$J_{\rm s} \approx -\frac{cH_0}{4\pi\lambda_{\rm L}^2}x$$

since $\cosh(x/\lambda_{\rm L}) \approx 1 + (1/2)(x/\lambda_{\rm L})^2 + \ldots$ and $\sinh(x/\lambda_{\rm L}) \approx x/\lambda_{\rm L} + \ldots$
Both cases $d \gg \lambda_{\rm L}$ and $d \ll \lambda_{\rm L}$ are shown in Figure S.3.67b.

APPENDIXES

Appendix 1
Approximate Values of Physical Constants

Constant	Symbol	SI	CGS
Speed of light	c	$3.00 \cdot 10^8$ m/s	$3.00 \cdot 10^{10}$ cm/s
Planck's constant	h	$6.63 \cdot 10^{-34}$ J $\cdot$ s	$6.63 \cdot 10^{-27}$ erg $\cdot$ s
Reduced Planck's constant	$\hbar = \dfrac{h}{2\pi}$	$1.05 \cdot 10^{-34}$ J $\cdot$ s	$1.05 \cdot 10^{-27}$ erg $\cdot$ s
Avogadro's number	N_A	$6.02 \cdot 10^{26}$ kmol^{-1}	$6.02 \cdot 10^{23}$ mol^{-1}
Boltzmann's constant	k	$1.38 \cdot 10^{-23}$ J/K	$1.38 \cdot 10^{-16}$ erg/K
Electron charge	e	$1.60 \cdot 10^{-19}$ C	$4.80 \cdot 10^{-10}$ esu
Electron mass	m_e	$9.11 \cdot 10^{-31}$ kg	$9.11 \cdot 10^{-28}$ g
Electron charge to mass ratio	$\dfrac{e}{m_e}$	$1.76 \cdot 10^{11}$ C/kg	$5.27 \cdot 10^{17}$ esu/g
Neutron mass	m_n	$1.675 \cdot 10^{-27}$ kg	$1.675 \cdot 10^{-24}$ g
Proton mass	m_p	$1.673 \cdot 10^{-27}$ kg	$1.673 \cdot 10^{-24}$ g
Gravitational constant	G	$6.67 \cdot 10^{-11}$ N $\cdot$ m^2/kg^2	$6.67 \cdot 10^{-8}$ dyn $\cdot$ cm^2/g^2
Acceleration of gravity	g	9.81 m/s^2	981 cm/s^2
Stefan-Boltzmann constant	σ	5.67 W/(m^{2}K^4)	$5.67 \cdot 10^{-5}$ erg/(s $\cdot$ cm^{2}K^4)
Fine structure constant	α	$1/137$	$1/137$
Bohr radius	a_0	$5.29 \cdot 10^{-11}$ m	$5.29 \cdot 10^{-9}$ cm
Classical electron radius	$r_e = \dfrac{e^2}{m_e c^2}$	$2.82 \cdot 10^{-15}$ m	$2.82 \cdot 10^{-13}$ cm
Electron Compton wavelength	$\dfrac{\hbar}{m_e c}$	$3.86 \cdot 10^{-13}$ m	$3.86 \cdot 10^{-11}$ cm
Bohr magneton	$\mu_B = \dfrac{e\hbar}{2m_e c}$	$9.27 \cdot 10^{-24}$ J/T	$9.27 \cdot 10^{-21}$ erg/G
Rydberg constant	R_∞	$1.10 \cdot 10^7$ m^{-1}	$1.10 \cdot 10^5$ cm^{-1}
Universal gas constant	R	$8.31 \cdot 10^3$ J/(kmol $\cdot$ K)	$8.31 \cdot 10^7$ erg/(mol $\cdot$ K)
Josephson constant	$\dfrac{2e}{h}$	$4.84 \cdot 10^{14}$ Hz/V	$1.45 \cdot 10^{17}$ Hz/statvolt
Permittivity of free space	ε_0	$8.85 \cdot 10^{-12}$ F/m	1

Some Astronomical Data

Mass of the Sun $M_S \approx 2 \cdot 10^{30}$ kg

Radius of the Sun $R_S \approx 6.7 \cdot 10^8$ m

Average Distance between the Earth and the Sun $\approx 1.5 \cdot 10^{11}$ m

Average Radius of the Earth $R_E \approx 6.4 \cdot 10^6$ m

Mass of the Earth $M_E \approx 6 \cdot 10^{24}$ kg

Average Velocity of the Earth in Orbit about the Sun $V_E \approx 3 \cdot 10^4$ m/s

Average Distance between the Earth and the Moon $\approx 3.8 \cdot 10^8$ m

Other Commonly Used Units

Angstrom $(\text{Å}) = 10^{-8}$ cm $= 10^{-10}$ m

Fermi $(\text{Fm}) = 10^{-13}$ cm $= 10^{-15}$ m

Barn $= 10^{-24}$ cm^2 $= 10^{-28}$ m^2

Year $\approx 3.16 \cdot 10^7$ s

Astronomical Year $\approx 9.5 \cdot 10^{17}$ cm $= 9.5 \cdot 10^{15}$ m

Parsec $= 3.1 \cdot 10^{18}$ cm $= 3.1 \cdot 10^{16}$ m

eV $= 1.6 \cdot 10^{-19}$ J $= 1.6 \cdot 10^{-12}$ erg

Room Temperature (294 K) ≈ 0.025 eV

Horsepower (hp) $= 746$ W

Calorie ≈ 4.2 J

Atmosphere $= 10^6$ dynes/cm^2

Appendix 2
Conversion Table from Rationalized MKSA to Gaussian Units

Physical Quantities	Rationalized MKSA	Conversion Coefficients	Gaussian
Charge	coulomb	$3 \cdot 10^9$	esu
Charge Density	coulomb/m^3	$3 \cdot 10^3$	esu/cm^3
Current	ampere	$3 \cdot 10^9$	esu/sec
Electric Field	volt/m	$\dfrac{1}{3 \cdot 10^4}$	statvolt/cm
Potential (Voltage)	volt	$1/300$	statvolt
Magnetic Flux	weber	10^8	gauss $\cdot$ cm^2 (maxwell)
Magnetic Induction	tesla	10^4	gauss
Magnetic Field	ampere-turn/m	$4\pi \cdot 10^{-4}$	oersted
Inductance	henry	$\dfrac{1}{9 \cdot 10^{11}}$	sec^2/cm
Capacitance	farad	$9 \cdot 10^{11}$	cm
Resistance	ohm	$\dfrac{1}{9 \cdot 10^{11}}$	sec/cm
Conductivity	mho/m	$9 \cdot 10^9$	sec^{-1}

Appendix 3

Vector Identities

$$\nabla \left(\Phi \Psi \right) = \Phi \nabla \Psi + \Psi \nabla \Phi$$

$$\nabla \cdot \left(\Phi \mathbf{A} \right) = \mathbf{A} \cdot \nabla \Phi + \Phi \nabla \cdot \mathbf{A}$$

$$\nabla \times \left(\Phi \mathbf{A} \right) = \Phi \nabla \times \mathbf{A} + \nabla \Phi \times \mathbf{A}$$

$$\nabla \cdot \left(\mathbf{A} \times \mathbf{B} \right) = \mathbf{B} \cdot \nabla \times \mathbf{A} - \mathbf{A} \cdot \nabla \times \mathbf{B}$$

$$\nabla \times \left(\mathbf{A} \times \mathbf{B} \right) = \left(\mathbf{B} \cdot \nabla \right) \mathbf{A} - \left(\mathbf{A} \cdot \nabla \right) \mathbf{B} + \mathbf{A} \left(\nabla \cdot \mathbf{B} \right) - \mathbf{B} \left(\nabla \cdot \mathbf{A} \right)$$

$$\left(\mathbf{A} \cdot \nabla \right) = A_x \frac{\partial}{\partial x} + A_y \frac{\partial}{\partial y} + A_z \frac{\partial}{\partial z}$$

$$\nabla \left(\mathbf{A} \cdot \mathbf{B} \right) = \left(\mathbf{B} \cdot \nabla \right) \mathbf{A} + \left(\mathbf{A} \cdot \nabla \right) \mathbf{B} + \mathbf{B} \times \left(\nabla \times \mathbf{A} \right) + \mathbf{A} \times \left(\nabla \times \mathbf{B} \right)$$

$$\nabla \cdot \nabla \Phi = \nabla^2 \Phi \equiv \Delta \Phi = \left(\frac{\partial^2}{\partial x^2} + \frac{\partial^2}{\partial y^2} + \frac{\partial^2}{\partial z^2} \right) \Phi$$

$$\nabla \times \nabla \times \mathbf{A} = \nabla \left(\nabla \cdot \mathbf{A} \right) - \left(\nabla^2 \right) \mathbf{A}$$

Vector Formulas

in Spherical and Cylindrical Coordinates

Spherical Coordinates

- **Transformation of Coordinates**

$$r = \sqrt{x^2 + y^2 + z^2}$$

$$\theta = \cos^{-1} \frac{z}{r}$$

$$\tan \varphi = \frac{y}{x}$$

$$x = r \sin \theta \cos \varphi$$

$$y = r \sin \theta \sin \varphi$$

$$z = r \cos \theta$$

- **Transformation of Differentials**

$$dx = \sin \theta \cos \varphi \; dr + r \cos \theta \cos \varphi \; d\theta - r \sin \theta \sin \varphi \; d\varphi$$

$$dy = \sin \theta \sin \varphi \; dr + r \cos \theta \sin \varphi \; d\theta + r \sin \theta \cos \varphi \; d\varphi$$

$$dz = \cos \theta \; dr - r \sin \theta \; d\theta$$

- **Square of the Element of Length**

$$ds^2 = dr^2 + r^2 \; d\theta^2 + r^2 \sin^2 \theta \; d\varphi^2$$

- **Transformation of the Coordinates of a Vector**

$$F_r = F_x \sin \theta \cos \varphi + F_y \sin \theta \sin \varphi + F_z \cos \theta$$

$$F_\theta = F_x \cos \theta \cos \varphi + F_y \cos \theta \sin \varphi - F_z \sin \theta$$

$$F_\varphi = -F_x \sin\varphi + F_y \cos\varphi$$

$$F_x = F_r \sin\theta \cos\varphi + F_\theta \cos\theta \cos\varphi - F_\varphi \sin\varphi$$

$$F_y = F_r \sin\theta \sin\varphi + F_\theta \cos\theta \sin\varphi + F_\varphi \cos\varphi$$

$$F_z = F_r \cos\theta - F_\theta \sin\theta$$

- **Divergence**

$$\nabla \cdot \mathbf{F} = \frac{\partial F_x}{\partial x} + \frac{\partial F_y}{\partial y} + \frac{\partial F_z}{\partial z}$$

$$= \frac{1}{r^2}\frac{\partial}{\partial r}\left(r^2 F_r\right) + \frac{1}{r\sin\theta}\frac{\partial}{\partial\theta}\left(F_\theta \sin\theta\right) + \frac{1}{r\sin\theta}\frac{\partial F_\varphi}{\partial\varphi}$$

- **Curl**

$$\nabla \times \mathbf{F} = \begin{vmatrix} \hat{\mathbf{x}} & \hat{\mathbf{y}} & \hat{\mathbf{z}} \\ \dfrac{\partial}{\partial x} & \dfrac{\partial}{\partial y} & \dfrac{\partial}{\partial z} \\ F_x & F_y & F_z \end{vmatrix}$$

$$= \frac{1}{r\sin\theta}\left[\frac{\partial}{\partial\theta}\left(F_\varphi \sin\theta\right) - \frac{\partial F_\theta}{\partial\varphi}\right]\mathbf{e}_r + \frac{1}{r}\left[\frac{1}{\sin\theta}\frac{\partial F_r}{\partial\varphi} - \frac{\partial\left(rF_\varphi\right)}{\partial r}\right]\mathbf{e}_\theta$$

$$+\frac{1}{r}\left[\frac{\partial\left(rF_\theta\right)}{\partial r} - \frac{\partial F_r}{\partial\theta}\right]\mathbf{e}_\varphi$$

- **Gradient**

$$\nabla\Phi = \frac{\partial\Phi}{\partial r}\mathbf{e}_r + \frac{1}{r}\frac{\partial\Phi}{\partial\theta}\mathbf{e}_\theta + \frac{1}{r\sin\theta}\frac{\partial\Phi}{\partial\varphi}\mathbf{e}_\varphi$$

- **Laplacian**

$$\nabla^2\Phi = \frac{\partial^2\Phi}{\partial x^2} + \frac{\partial^2\Phi}{\partial y^2} + \frac{\partial^2\Phi}{\partial z^2}$$

$$= \frac{1}{r^2}\frac{\partial}{\partial r}\left(r^2\frac{\partial\Phi}{\partial r}\right) + \frac{1}{r^2\sin\theta}\frac{\partial}{\partial\theta}\left(\sin\theta\frac{\partial\Phi}{\partial\theta}\right) + \frac{1}{r^2\sin^2\theta}\frac{\partial^2\Phi}{\partial\varphi^2}$$

Cylindrical Coordinates

- **Transformation of Coordinates**

$$\rho = \sqrt{x^2 + y^2}$$

$$\tan\varphi = \frac{y}{x}$$

$$z = z$$

$$x = \rho\cos\varphi$$

$$y = \rho\sin\varphi$$

$$z = z$$

- **Transformation of Differentials**

$$dx = \cos\varphi\, d\rho - \rho\sin\varphi\, d\varphi$$

$$dy = \sin\varphi\, d\rho + \rho\cos\varphi\, d\varphi$$

$$dz = dz$$

- **Square of the Element of Length**

$$ds^2 = d\rho^2 + \rho^2\, d\varphi^2 + dz^2$$

- **Transformation of the Coordinates of a Vector**

$$F_\rho = F_x\cos\varphi + F_y\sin\varphi$$

$$F_\varphi = -F_x\sin\varphi + F_y\cos\varphi$$

$$F_z = F_z$$

$$F_x = F_\rho \cos\varphi - F_\varphi \sin\varphi$$

$$F_y = F_\rho \sin\varphi + F_\varphi \cos\varphi$$

$$F_z = F_z$$

- Divergence
$$\nabla \cdot \mathbf{F} = \frac{1}{\rho}\frac{\partial\left(\rho F_\rho\right)}{\partial\rho} + \frac{1}{\rho}\frac{\partial F_\varphi}{\partial\varphi} + \frac{\partial F_z}{\partial z}$$

- Curl
$$\nabla \times \mathbf{F} = \left(\frac{1}{\rho}\frac{\partial F_z}{\partial\varphi} - \frac{\partial F_\varphi}{\partial z}\right)\mathbf{e}_\rho + \left(\frac{\partial F_\rho}{\partial z} - \frac{\partial F_z}{\partial\rho}\right)\mathbf{e}_\varphi + \frac{1}{\rho}\left(\frac{\partial\left(\rho F_\varphi\right)}{\partial\rho} - \frac{\partial F_\rho}{\partial\varphi}\right)\hat{\mathbf{z}}$$

- Gradient
$$\nabla\Phi = \frac{\partial\Phi}{\partial\rho}\mathbf{e}_\rho + \frac{1}{\rho}\frac{\partial\Phi}{\partial\varphi}\mathbf{e}_\varphi + \frac{\partial\Phi}{\partial z}\hat{\mathbf{z}}$$

- Laplacian
$$\nabla^2\Phi = \frac{1}{\rho}\frac{\partial}{\partial\rho}\left(\rho\frac{\partial\Phi}{\partial\rho}\right) + \frac{1}{\rho^2}\frac{\partial^2\Phi}{\partial\varphi^2} + \frac{\partial^2\Phi}{\partial z^2}$$

Appendix 4

Legendre Polynomials

$$P_0(x) = 1$$

$$P_1(x) = x$$

$$P_2(x) = \frac{1}{2}\left(3x^2 - 1\right)$$

$$P_3(x) = \frac{1}{2}\left(5x^3 - 3x\right)$$

$$P_4(x) = \frac{1}{8}\left(35x^4 - 30x^2 + 3\right)$$

Rodrigues' Formula

$$P_l(x) = \frac{1}{2^l l!} \frac{d^l}{dx^l} \left(x^2 - 1\right)^l$$

Spherical Harmonics

$$Y_0^0(\theta, \varphi) = \frac{1}{\sqrt{4\pi}}$$

$$Y_1^1(\theta, \varphi) = -\sqrt{\frac{3}{8\pi}}\ \sin\theta\ e^{i\varphi}$$

$$Y_1^0(\theta, \varphi) = \sqrt{\frac{3}{4\pi}}\ \cos\theta$$

$$Y_1^{-1}(\theta, \varphi) = \sqrt{\frac{3}{8\pi}}\ \sin\theta\ e^{-i\varphi}$$

$$Y_2^2(\theta, \varphi) = \frac{1}{4}\sqrt{\frac{15}{2\pi}}\ \sin^2\theta\ e^{2i\varphi}$$

$$Y_2^1(\theta, \varphi) = -\sqrt{\frac{15}{8\pi}}\ \sin\theta\cos\theta\ e^{i\varphi}$$

$$Y_2^0(\theta, \varphi) = \frac{1}{2}\sqrt{\frac{5}{4\pi}}\left(3\cos^2\theta - 1\right)$$

$$Y_2^{-1}(\theta, \varphi) = \sqrt{\frac{15}{8\pi}}\ \sin\theta\cos\theta\ e^{-i\varphi}$$

$$Y_2^{-2}(\theta, \varphi) = \frac{1}{4}\sqrt{\frac{15}{2\pi}}\ \sin^2\theta\ e^{-2i\varphi}$$

Bibliography

Arfken, G., *Mathematical Methods for Physicists*, 3rd ed., Orlando: Academic Press, 1985

Arnold, V. I., *Mathematical Methods of Classical Mechanics*, 2nd ed., New York: Springer-Verlag, 1978

Barger, V.D., and Olsson, M.G., *Classical Mechanics, A Modern Perspective*, New York: McGraw-Hill, 1973

Batygin, V.V., and Toptygin, I.N., *Problems in Electrodynamics*, 2nd ed., London: Academic Press, 1978

Chen, M., *University of California, Berkeley, Physics Problems, with Solutions*, Englewood Cliffs, NJ: Prentice-Hall, Inc., 1974

Cronin, J., Greenberg, D., and Telegdi, V., *University of Chicago Graduate Problems in Physics*, Chicago, University of Chicago Press, 1979

Feynman, R., Leighton, R., and Sands, M., *The Feynman Lectures on Physics*, Reading, MA, Addison-Wesley, 1965

Goldstein, H., *Classical Mechanics*, 2nd ed., Reading, MA: Addison-Wesley, 1981

Halzen, F., and Martin, A., *Quarks and Leptons*, New York: John Wiley & Sons, Inc., 1984

Jackson, J.D., *Classical Electrodynamics*, New York: John Wiley & Sons, Inc., 1975

Johnson, W., *Helicopter Theory*, Princeton: Princeton University Press, 1980

Kittel, C., *Introduction to Solid State Physics*, 6th ed., New York: John Wiley & Sons, Inc., 1986

Kober, H., *Dictionary of Conformal Representations*, New York: Dover, 1957

Kozel, S.M., Rashba, E.I., and Slavatinskii, S.A., *Problems of the Moscow Physico-Technical Institute*, Moscow: Mir, 1986

Landau, L.D., and Lifshitz, E.M., *Mechanics,* Volume 1 of *Course of Theoretical Physics,* 3rd ed., Elmsford, New York: Pergamon Press, 1976

Landau, L.D., and Lifshitz, E.M., *Classical Theory of Fields,* Volume 2 of *Course of Theoretical Physics,* 4th ed., Elmsford, New York: Pergamon Press, 1975

Landau, L.D., and Lifshitz, E.M., *Fluid Mechanics,* Volume 6 of *Course of Theoretical Physics,* 2nd ed., Elmsford, New York: Pergamon Press, 1987

Landau, L.D., and Lifshitz, E.M., and Pitaevskiĭ, L.P., *Electrodynamics of Continuous Media,* Volume 8 of *Course of Theoretical Physics,* 2nd ed., Elmsford, New York: Pergamon Press, 1984

Marion, J.B., and Heald, M.A., *Classical Electromagnetic Radiation*, 2nd ed., New York: Academic Press, 1980

Newbury, N., Newman, M., Ruhl, J., Staggs, S., and Thorsett, S., *Princeton Problems in Physics, with Solutions*, Princeton: Princeton University Press, 1991

Panofsky, W., and Philips, M., *Classical Electricity and Magnetism*, 2nd ed., Reading, MA: Addison-Wesley, 1962

Purcell, E.M., *Electricity and Magnetism,* Volume 2 of *Berkeley Physics Course*, New York: McGraw-Hill, 1965

Routh, E., *Dynamics of a System of Rigid Bodies*, New York: Dover, 1960

Smythe, W.R., *Static and Dynamic Electricity,* 3rd ed., New York: Hemisphere Publishing Corp., 1989

Spiegel, M., *Schaum's Outline, Theory and Problems of Complex Variables*, New York: McGraw-Hill, 1964

Taylor, E.F., and Wheeler, J.A., *Spacetime Physics*, San Francisco, California: W.H. Freeman and Company, 1966

Tinkham, M., *Introduction to Superconductivity*, New York, McGraw-Hill, 1975